VOLUME THIRTY TWO

ADVANCES IN BIOMEMBRANES AND LIPID SELF-ASSEMBLY

EDITORIAL BOARD

VOLUME THIRTY TWO

ADVANCES IN BIOMEMBRANES AND LIPID SELF-ASSEMBLY

Biomembrane Vesicles: Scientific, Clinical and Technological Considerations

Edited by

Antonella Bongiovanni
Institute for Biomedical Research and Innovation (IRIB)
National Research Council of Italy (CNR), Palermo, Italy

Gabriella Pocsfalvi
Extracellular Vesicles and Mass Spectrometry Laboratory, Institute of Biosciences and BioResources (IBBR), National Research Council of Italy (CNR), Naples, Italy

Mauro Manno
Institute of Biophysics (IBF)
National Research Council of Italy (CNR), Palermo, Italy

Veronika Kralj-Iglič
Laboratory of Clinical Biophysics, Faculty of Health Sciences
University of Ljubljana, Ljubljana, Slovenia
Extracellular Vesicles and Mass Spectrometry Laboratory, Institute of Biosciences and BioResources (IBBR), National Research Council of Italy (CNR), Naples, Italy

ELSEVIER

ACADEMIC PRESS
An imprint of Elsevier

Academic Press is an imprint of Elsevier
125 London Wall, London EC2Y 5AS, United Kingdom
525 B Street, Suite 1650, San Diego, CA 92101, United States
50 Hampshire Street, 5th Floor, Cambridge, MA 02139, United States
The Boulevard, Langford Lane, Kidlington, Oxford OX5 1GB, United Kingdom

First edition 2020

ISBN: 978-0-12-820968-4
ISSN: 2451-9634

For information on all Academic Press publications visit our website at https://www.elsevier.com/books-and-journals

Publisher: Zoe Kruze
Acquisitions Editor: Jason Mitchell
Editorial Project Manager: Shellie Bryant
Production Project Manager: Abdulla Sait
Cover Designer: Greg Harris

Typeset by TNQ Technologies

Contents

Contributors

Giorgia Adamo
Institute for Biomedical Research and Innovation (IRIB), National Research Council of Italy (CNR), Palermo, Italy

Heinz Amenitsch
Graz University of Technology, Graz, Austria

Antonella Bongiovanni
Institute for Biomedical Research and Innovation (IRIB), National Research Council of Italy (CNR), Palermo, Italy

Darja Božič
Laboratory of Clinical Biophysics, Faculty of Health Sciences, University of Ljubljana, Ljubljana, Slovenia; Laboratory of Physics, Faculty of Electrical Engineering, University of Ljubljana, Ljubljana, Slovenia

Valentina Buffa
Institute for Biomedical Research and Innovation (IRIB), National Research Council of Italy (CNR), Palermo, Italy; Integrare UMR_S951Genethon, Inserm, University of Evry, Université Paris Saclay Genethon, Evry, France

Giovambattista Capasso
First Division of Nephrology, Department of Clinical and Experimental Medicine, School of Medicine, University of Campania Luigi Vanvitelli, Naples, Italy

Federica Caponnetto
Università Degli Studi di Udine, Udine, Italy

Rita Carrotta
CNR, Istituto di Biofisica, Palermo, Italy

Loredana Casalis
Elettra Sincrotrone Trieste S.C.p.A., Trieste, Italy

Daniela Cesselli
Università Degli Studi di Udine, Udine, Italy

Antonella Cusimano
Institute for Biomedical Research and Innovation (IRIB), National Research Council of Italy (CNR), Palermo, Italy

Damjana Drobne
Department of Biology, Biotechnical Faculty, University of Ljubljana, Ljubljana, Slovenia

Francesco D'Amico
Elettra Sincrotrone Trieste S.C.p.A., Trieste, Italy

Immacolata Fiume
Extracellular Vesicles and Mass Spectrometry Laboratory, Institute of Biosciences and BioResources (IBBR), National Research Council of Italy (CNR), Naples, Italy

Mario Gimona
GMP Unit, Spinal Cord Injury and Tissue Regeneration Center Salzburg, Paracelsus Medical University, Salzburg, Austria; Research Program "Nanovesicular Therapies", Paracelsus Medical University, Salzburg (Austria)

Matej Hočevar
Department of Physics and Chemistry of Materials, Institute of Metals and Technology, Ljubljana, Slovenia

Aleš Iglič
Laboratory of Physics, Faculty of Electrical Engineering, University of Ljubljana, Ljubljana, Slovenia; Extracellular Vesicles and Mass Spectrometry Laboratory, Institute of Biosciences and BioResources (IBBR), National Research Council of Italy (CNR), Naples, Italy; Laboratory of Clinical Biophysics, Faculty of Medicine, University of Ljubljana, Ljubljana, Slovenia

Marko Jeran
Laboratory of Clinical Biophysics, Faculty of Health Sciences, University of Ljubljana, Ljubljana, Slovenia; Laboratory of Physics, Faculty of Electrical Engineering, University of Ljubljana, Ljubljana, Slovenia

Ksenija Kogej
Department of Chemistry and Biochemistry, Faculty of Chemistry and Chemical Technology, University of Ljubljana, Ljubljana, Slovenia

Veno Kononenko
Department of Biology, Biotechnical Faculty, University of Ljubljana, Ljubljana, Slovenia

Veronika Kralj-Iglič
Laboratory of Clinical Biophysics, Faculty of Health Sciences, University of Ljubljana, Ljubljana, Slovenia; Extracellular Vesicles and Mass Spectrometry Laboratory, Institute of Biosciences and BioResources (IBBR), National Research Council of Italy (CNR), Naples, Italy

Fabio Librizzi
CNR, Istituto di Biofisica, Palermo, Italy

Maria Rosalia Mangione
CNR, Istituto di Biofisica, Palermo, Italy

Magdalena Mayr
GMP Unit, Spinal Cord Injury and Tissue Regeneration Center Salzburg, Paracelsus Medical University, Salzburg, Austria

Oscar Moran
CNR, Istituto di Biofisica, Genova, Italy

Maria Pachetti
Elettra Sincrotrone Trieste S.C.p.A., Trieste, Italy; Università Degli Studi di Trieste, Trieste, Italy

Karin Pachler
GMP Unit, Spinal Cord Injury and Tissue Regeneration Center Salzburg, Paracelsus Medical University, Salzburg, Austria; Research Program "Nanovesicular Therapies", Paracelsus Medical University, Salzburg (Austria)

Ljubiša Pađen
Laboratory of Clinical Biophysics, Faculty of Health Sciences, University of Ljubljana, Ljubljana, Slovenia

Manca Pajnič
Laboratory of Clinical Biophysics, Faculty of Health Sciences, University of Ljubljana, Ljubljana, Slovenia

Pietro Parisse
Elettra Sincrotrone Trieste S.C.p.A., Trieste, Italy

Fabio Perissinotto
Elettra Sincrotrone Trieste S.C.p.A., Trieste, Italy

Sabrina Picciotto
Institute for Biomedical Research and Innovation (IRIB), National Research Council of Italy (CNR), Palermo, Italy

Gabriella Pocsfalvi
Extracellular Vesicles and Mass Spectrometry Laboratory, Institute of Biosciences and BioResources (IBBR), National Research Council of Italy (CNR), Naples, Italy

Nataša Poklar Ulrih
Department of Food Science and Technology, Biotechnical Faculty, University of Ljubljana, Ljubljana, Slovenia

Ana Paulina Ramos Juarez
Extracellular Vesicles and Mass Spectrometry Laboratory, Institute of Biosciences and BioResources (IBBR), National Research Council of Italy (CNR), Naples, Italy; First Division of Nephrology, Department of Clinical and Experimental Medicine, School of Medicine, University of Campania Luigi Vanvitelli, Naples, Italy

Eva Rohde
GMP Unit, Spinal Cord Injury and Tissue Regeneration Center Salzburg, Paracelsus Medical University, Salzburg, Austria; Department of Transfusion Medicine, University Hospital, Salzburger Landeskliniken Ges.m.b.H (SALK), Salzburg, Austria

Daniele P. Romancino
Institute for Biomedical Research and Innovation (IRIB), National Research Council of Italy (CNR), Palermo, Italy

Barbara Sartori
Graz University of Technology, Graz, Austria

Beatrice Senigagliesi
Elettra Sincrotrone Trieste S.C.p.A., Trieste, Italy; Università Degli Studi di Trieste, Trieste, Italy

Urška Štibler
Laboratory of Physics, Faculty of Electrical Engineering, University of Ljubljana, Ljubljana, Slovenia

Luka Šturm
Department of Food Science and Technology, Biotechnical Faculty, University of Ljubljana, Ljubljana, Slovenia

Francesco Trepiccione
First Division of Nephrology, Department of Clinical and Experimental Medicine, School of Medicine, University of Campania Luigi Vanvitelli, Naples, Italy

Lisa Vaccari
Elettra Sincrotrone Trieste S.C.p.A., Trieste, Italy

Silvia Vilasi
CNR, Istituto di Biofisica, Palermo, Italy

Preface

Throughout the last decade, our understanding of cell-to-cell communication has made an exceptional progress, which in turn has revealed biological membranes as key players in cell life and development. The finding that cell-derived membranous nanoparticles (e.g., extracellular vesicles, EVs) mediate intercellular communication by transferring biomolecules, including proteins, lipids, DNA and RNA has revolutionized cell biology and epigenetic fields, opening new prospects for the cross-talk between cells within the same organ, and among different tissues and even organisms in health and disease and as cross-kingdom communication mean.

Furthermore, the discovery of EVs as carriers of functional small molecules and proteins has raised great interest in the drug delivery field, as it may be possible to harness these vesicles for therapeutic delivery of miRNA, siRNA, mRNA, lncRNA, peptides and synthetic drugs. Native and drug-loaded mammalian cell-derived EVs (e.g., mesenchymal cell-derived EVs) have been recently identified and now constitute a rapidly growing research interest known as "cell-free therapy".

In this special volume, original research articles and reviews on all aspects related to the theme of "Extracellular Vesicles: from basic sciences to therapeutic exploitation" are presented. The volume features different functional and applicative themes revealing a fundamental role of EVs in membrane biology, biochemistry and biophysics, and in the functioning of living organisms including human beings. The contributions are organized into two volumes with 6 chapters in each volume.

In Volume 32, Chapters 1 and 2 (Amyloid β-peptide interaction with GM1 containing model membrane by Rita Carrotta et al., and Propolis flavonoids and terpenes, and their interactions with the model lipid membranes: a review, by Luka Šturm and Nataša Poklar Ulrich, respectively) consider molecular interactions that underlie membrane composition and function. Protein composition, regulation and their relation to the EV biological functions are presented in Chapters 3 and 4 (The Human EV membranome by Ana Paulina Ramos Juarez et al., and Post-translational lipidation in extracellular vesicles: chemical mechanism, biological function and application by Sabrina Picciotto et al., respectively). Close connection between methods and identity of EVs are addressed in Chapters 5 and 6 (Pursuing mechanisms of extracellular vesicle formation. Effects of sample processing, by Darja Božič et al., and Multi-technique analysis of extracellular vesicles: not only size matters by Fabio Perissinotto et al., respectively).

In Volume 33, the clinical aspects of extracellular vesicles are considered in Chapters 1–3 (Single patient analysis for clinical applications by Christopher Stanly et al., Biological role and clinical relevance of extracellular vesicles as key mediators of cell communication in cancer by Francesca Mantile et al., and Treatment with platelet- and extracellular vesicle-rich plasma in otorhinolaryngology. A review and future perspectives by Domen Vozel et al., respectively). Chapters 4 and 5 provide important insights into future development of the field of EVs. Plant EVs as novel and promising systems for industrial application are introduced in Chapter 4 (Plant extracellular vesicles: the safe for bioactive compounds by Stefania Raimondo and Marzia Pucci) while the problem of repeatability and accuracy of measurement that is common to all research in EV field is presented in Chapter 5 (A quality management based approach for exosomes research by Annamaria Kisslinger and Giovanna Liguori). The last chapter completes the volume with insights into the evolution of membrane systems from which we may learn fundamental mechanisms connecting living and non-living systems (Evolution of mechanical stability from lipid layers to complex bacterial envelope structures by Žiga Pandur and David Stopar).

CHAPTER ONE

Amyloid β-peptide interaction with GM1 containing model membrane

Rita Carrotta[a,*], Fabio Librizzi[a], Maria Rosalia Mangione[a], Silvia Vilasi[a] and Oscar Moran[b]
[a]CNR, Istituto di Biofisica, Palermo, Italy
[b]CNR, Istituto di Biofisica, Genova, Italy
[*]Corresponding author: E-mail: rita.carrotta@cnr.it

Contents

Abstract

Early impairments in Alzheimer's disease are related to neuronal membrane damage. Aberrant Amyloid β (Aβ) species and specific membrane components both play a role in promoting aggregation, deposition and signal dysfunction. Ganglioside GM1, present with cholesterol and sphingomyelin in lipid rafts, seems to be able to interact with Aβ and initiate Aβ aggregation on membrane. In this chapter, several studies are reviewed, highlighting the role of different lipids and in particular GM1 in the interaction between Aβ and membrane systems. Results concerning GM1-Aβ interaction obtained by means of different experimental methods are reported and discussed. Small angle X-ray scattering is used to study the structure of the unilamellar vesicle bilayer. Mixing Aβ species with vesicles clearly showed an effect on the bilayer structure, but only for GM1-containing liposomes. The elastic properties of GM1-containing

Advances in Biomembranes and Lipid Self-Assembly, Volume 32
ISSN 2451-9634
https://doi.org/10.1016/bs.abl.2020.04.001

membrane have been studied by Neutron Spin Echo, showing that the presence of Aβ species brings about an increase in the membrane rigidity, as reflected by the measured elastic modulus. This effect can be hindered by the presence of molecular chaperones, like Hsp60. This chaperonin has proven to be a strong inhibitor of the amyloidogenesis process, due to its dose dependent ability to neutralize oligomeric metastable species needed to onset the fibril growth cascade. The fact that the incubation of Aβ in the presence of Hsp60 leads to a loss of the Aβ effects on the membrane dynamical properties suggests that the aberrant interaction between Aβ and membranes is important in these specific intermediate Aβ species.

List of abbreviations

Aβ Amyloid β-peptide
GM1 Monosialotetrahexosyl ganglioside
APP Amyloid Precursor Protein
Hsp60 heat shock protein 60
POPC 1-palmitoyl-2-oleoyl-sn-glycero-3-phosphocholine
POPS 1-palmitoyl-2-oleoyl-sn-glycero-3-phospho-L-serine
Chol Cholesterol
SM Sphingomyelin
DOPC 1,2-Dioleoyl-sn-glycero-3-phosphocholine
DSPC 1,2-distearoyl-sn-glycero-3-phosphocholine

1. Introduction

Nowadays, Alzheimer's disease (AD) etiology still remains an unsolved issue. Beside the evident neurodegenerative symptoms, a pathological hallmark of the disease is the formation of extracellular and intracellular protein aggregates, respectively composed of amyloid β (Aβ) peptide and tau protein [1−3]. However, how well the amount of proteinaceous deposits correlates with the progression of the disease still remains debated. In spite of an intensive effort, no diagnostic markers have been found yet to detect the disease at early stages. In the last 40 years research has linked many disorders to protein misfolding and aggregation; these disorders have been grouped as protein misfolding diseases, and AD is one of those [4−6]. The idea that small oligomeric species form in vivo, by escaping the protein quality control system in the cell, explains the difficulty to clearly detect the early stages of these diseases [7,8]. Aβ-peptides originate from processing of the trans membrane Amyloid Precursor Protein (APP) through secretase enzymes which yield peptides of different lengths. These peptides are released in the extracellular medium and have different propensities to aggregate [9]. Although plaques deposition observed in the neuronal extracellular space has been considered the main hallmark of AD, there is also evidence of Aβ intracellular accumulation, which is considered responsible of early synaptic dysfunction and memory impairment [10,11]. The presence of intracellular Aβ is sustained by the fact that APP can be found in the neuronal plasma membrane as well as in other intracellular compartment membranes, and that also secretases are present in different subcellular milieu. Moreover, some studies demonstrate that Aβ can be internalized from the extracellular space [11]. Aβ has been found in plaques together with a sphingolipid present in membranes, the ganglioside GM1 [12]. This suggested a possible role played by this glycolipid in the interaction with Aβ-peptides. Membranes are not homogeneous and are enriched of many components such as proteins that need a lipid surrounding specific for their structure-dynamic-function, to undertake conformational changes for molecular uptaking, channeling or enzymatic reactions. In neuron membranes, regions rich of cholesterol, GM1 and sphingomyelin (so called lipid rafts that exhibit high ordering), are the preferred regions for signaling pathways. The interaction of Aβ with membrane has been considered one of the possible causes of early neurologic disfunction. Small species of Aβ, internalized or produced in the interior of the cell, can imply some specific perturbation to the lipid matrix from a physico-chemical-structural point of view and interfere with some receptors function. This chapter concerns the interaction of Aβ-peptide with the ganglioside GM1. Unilamellar vesicles as model membrane systems, made with selected lipid components, allow to study the role of selected components in the matrix, from a structural and dynamical point of view. Small angle X-ray scattering (SAXS) is used to detect changes in the bilayer structure, while Neutron Spin Echo (NSE) measurements allow to monitor differences in the membrane stiffness due to the interaction between the lipid matrix and the Aβ peptide. A review of the literature is made, focusing on the bilayer-Aβ interaction upon different regimes of sphingolipid:lipid concentration ratio. In fact, different results are reported, depending on GM1 concentration, sphingomyelin presence, lipid composition and phase.

2. Large unilamellar vesicles

One of the most powerful strategies to investigate the role of membrane and in particular of the lipid composition of the bilayer in the interaction with Aβ-peptide species is the use of model systems, as for instance unilamellar vesicles (UV). These synthetic models can be prepared by using different methods, depending on the desired dimensions. UV are divided in small unilamelar vesicles (SUV), large unilamelar vesicles (LUV) and giant unilamelar vesicles (GUV) depending on their dimensions, in the range of 20—100 nm (SUV), 100—1000 nm (LUV) and larger, up to ten microns (GUV). By thin layer evaporation, a lipid film is built from a lipid organic solution. The lipid components are mixed in the chosen ratio and constitute the film. Hydration of the film leads to a spontaneous, heterogeneous solution of multilamellar lipid vesicles (MLV), representing the lowest free energy minimum of the hydrated system. The unilamelar structure has a higher free energy and therefore the system needs energy to jump to this metastable configuration [13].

As a model system, GUV offer the advantage to be large enough to be visible by light microscopy, although suffering poor stability. Among unilamelar structures, LUV present the highest stability. The results reported in this chapter focus on LUV as a model to study the interaction with the Aβ-peptide. LUV can be made up of suitably chosen lipid compositions by using methods like sonication or mechanical extrusion through a filter. Phospholipid charge, carbon tail unsaturation or length, presence of cholesterol or addition of specific components, such as sphingolipids, determine the structural and thermodynamic characteristics of the bilayer. Characterization of the vesicles can be achieved by evaluating concentration, dimension, bilayer width and elasticity properties. Many different methods are used to collect all the information needed to have a complete characterization such as Light Scattering, Small Angle X-ray Scattering and Neutron Spin Echo.

2.1 Light scattering

Light scattering (LS) detects the presence of vesicles in solution at very low concentration. The elastic static scattering intensity, the so-called Rayleigh ratio, is sensitive to the particle molecular weight and to the mass concentration of the nanoparticles, the particle scattering power being proportional to the squared differential refractive index of lipids with respect to the solvent. The lipid concentration constitutes the mass concentration of the solution. By modeling a phosphatidylcholine (PC) unilamellar vesicle of 100 nm diameter (bilayer width of about 4 nm) and using Area/PC = 0.63 nm^2 [14] for the polar head occupied by a PC molecule, the ratio between the lipid molar concentration and the vesicle number concentration is about 10^5. Therefore, the mass of such a vesicle will be about

$M_{Ves} = 10^5$ PC molecules, thus giving a high scattering signal. By using multi-angle light scattering (MALS), the scattering intensity $I(Q)$ as a function of the exchanged wave vector $Q = 4\pi/\lambda \sin(\theta/2)$ can be measured. From $I(Q) = I(0)P(Q)$, where $P(Q)$ is the form factor of the nanoparticles, it is possible to obtain the intensity averaged gyration radius by means of a simple Guinier analysis, or even information on the nanoparticle mass distribution by using more elaborate modeling. An unusual application of static LS is the concentration control of a LUV sample made by the extrusion method.

2.2 LUV mass concentration by static light scattering (SLS)

In order to appreciate a possible loss of mass during the extrusion step in the LUV preparation, the final concentration of an extruded solution can be estimated by assuming the starting lipid mass and using LS [15]. After having performed a first vesicle extrusion (step 1), the recovery of the whole lipid mass from the apparatus is achieved by adding buffer to perform a second rinsing extrusion (step 2), by obtaining two samples, corresponding to the two extrusion steps, S_1 and S_2. In order to obtain an estimate of the lipid concentration in S_1 and S_2, the scattering intensities, I_1 and I_2, are measured at a fixed angle (usually 90°), and their ratio I_1/I_2 is calculated. Appropriate dilution (d_1 and d_2 respectively) is applied. Since the extruded volumes (V_1 and V_2) as well as the total amount of solution used in the preparation and the dead volume (V_t and $\Delta V = V_t - V_1 - V_2$ respectively) can be measured during the extrusions, the lipid concentration in the main sample, coming from the extrusion 1 can be calculated as $c_1 = M/V_1\ [(1 + ((I_2\ d_2)/(I_1\ d_1))(V_2 + \Delta V)/V_1]^{-1}$, where M is the known lipid mass.

LUV size characterization by dynamic light scattering (DLS) measures the intensity correlation function, $g_2(t)$, and allows to estimate the hydrodynamic dimensions of the vesicles. From $g_2(t)$, the field autocorrelation, $g_1(t)$, is obtained by $g_1(t) = [g_2(t) - 1]^{1/2}$, and analyzed by using a smoothing constrained regularization method [16], in order to determine the distribution of relaxation times according to:

$$g_1(t) = \int A(\Gamma)\exp(-\Gamma t)d\Gamma \qquad (1)$$

where $A(\Gamma)$ denotes the contribution amplitude of the mode with characteristic time Γ^{-1}. The latter is related to the diffusion coefficient by $\Gamma = D\ q^2$. $A(\Gamma)$ can be expressed also as $A(R_h)$, where R_h is the hydrodynamic radius, by using the Stokes-Einstein relationship $D = K_B T/6\pi\eta R_h$, where η

is the viscosity, and K_B *and* T are the Boltzmann constant and the absolute temperature, respectively. The distribution obtained by data analysis is intensity weighted. To obtain number distribution, the vesicle mass can be considered proportional to R_h^2, a reasonable assumption for LUV geometry where the mass is distributed on an equivalent external shell (see Fig. 1, adapted from Ref. [15]). The harmonic intensity average radius can be obtained from the weighted intensity distribution.

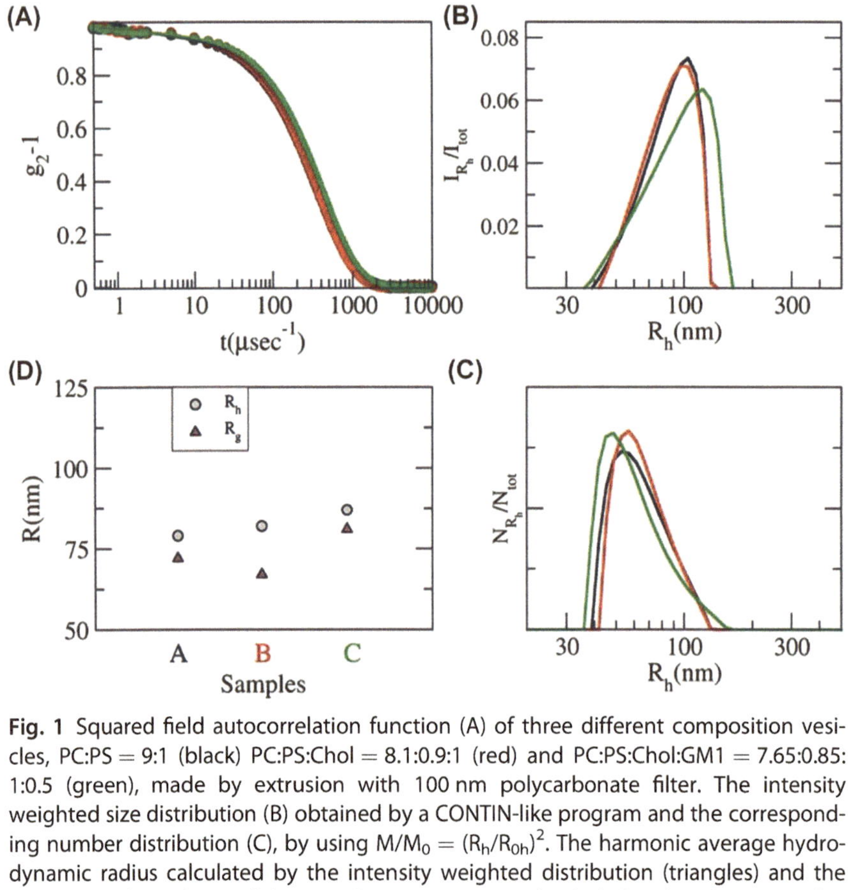

Fig. 1 Squared field autocorrelation function (A) of three different composition vesicles, PC:PS = 9:1 (black) PC:PS:Chol = 8.1:0.9:1 (red) and PC:PS:Chol:GM1 = 7.65:0.85:1:0.5 (green), made by extrusion with 100 nm polycarbonate filter. The intensity weighted size distribution (B) obtained by a CONTIN-like program and the corresponding number distribution (C), by using $M/M_0 = (R_h/R_{0h})^2$. The harmonic average hydrodynamic radius calculated by the intensity weighted distribution (triangles) and the gyration radius obtained by MALS measurements (circles) for the three vesicles. *Reprinted with permission from Nicastro et al. Biophys Chem 208 (2016) 9–16. Copyright 2016 Elsevier M.C. Nicastro, D. Spigolon, F. Librizzi, O. Moran, M.G. Ortore, D. Bulone, P.L. San Biagio, R. Carrotta, Amyloid β-peptide insertion in liposomes containing GM1-cholesterol domains, Biophys. Chem. 208 (2016) 9–16.*

2.3 Small angle X-ray scattering

As visible radiation is able to measure particle features in the range of 50 nm–5 μm and the form factor hides information about the shape or mass density distribution of a vesicle, X-rays interference comes from structures with characteristic length scales on the order of few nanometers, such as the leaflet dimension of the bilayer. X-rays (with radiation $\lambda < 0.2$ nm) are elastically scattered at the electrons of the atomic shell. Each atom has a characteristic X-ray scattering factor, which depends on its electronic configuration. In a particle built by a number of molecules with particular atomic composition the X-ray scattering is obtained by subtracting from the particle electronic density the average contribution of the environment. This spatial function is called contrast.

According to Debye scattering theory [17] the measured X-ray intensity is an average over a polydisperse vesicle population, so that

$$\langle I(Q)\rangle = N\langle F(Q)^2\rangle + \langle F(Q)^2\rangle \sum_{n,n'}^{n} \langle \cos(Q\cdot r_{nn'})\rangle \tag{2}$$

where $F(Q)$ is the bilayer form factor, N is the number of particles (vesicles) and $r_{nn'}$ is the inter-vesicle vector. The first term of Eq. (2) represents the average scattering of N individual vesicles, while the second term comes from the interference between vesicles and depends on the distance between them, $r_{nn'}$. In dilute conditions no interference between vesicles is expected [17]. In fact, the scattering coming from the second term becomes significant at $Q << 0.1\ \text{nm}^{-1}$, a regime of X-ray called USAXS [18]. When inter vesicle interactions can be neglected, the relation (2) becomes:

$$\langle I(Q)\rangle \alpha \langle F(Q)^2\rangle \tag{3}$$

where N is included in the arbitrary instrumental scaling. The form factor, $F(Q)$, is the Fourier transform of the electronic density $\rho(r)$ of the bilayer. To calculate the electronic density, a model of the membrane needs to be constructed, and the parameters of the model adjusted to fit its Fourier transform to the experimental SAXS spectra. The electronic density of the vesicle wall can be described by five concentric Gaussian shells [19–23], that compose an asymmetric bilayer profile with added decorations on the inner and outer sides of the vesicle wall as shown in Fig. 2. The bilayer electronic density profile is modeled by two positive Gaussian distributed layers, representing the head-groups of the two lipid leaflets (*in* and *out*, respectively) and a negative Gaussian distributed layer representing the hydrophobic core

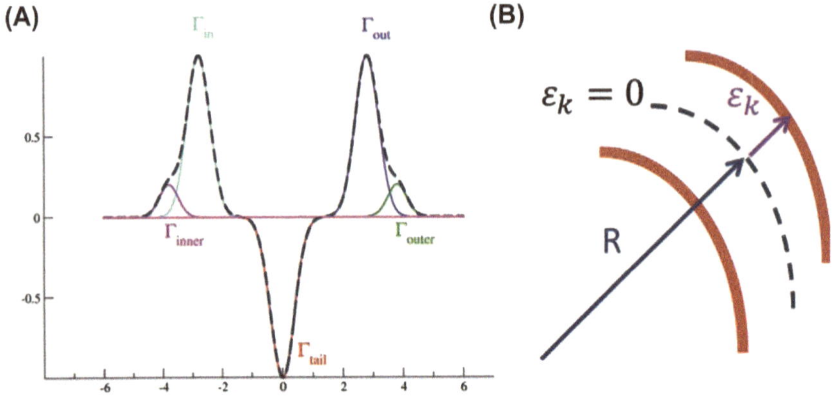

Fig. 2 Gaussian electronic density modeling profile (A); Scheme for clear visualization of the fitting parameters (B).

(*tail*). The asymmetrical distributed material extending beyond the bilayer toward the internal or external sides of the vesicle (*inner* and *outer*) can be modeled by a Gaussian electronic distribution, in the inner and outer sides of the bilayer, respectively. The electronic density of the bilayer profile can be then written:

$$\rho(r) = \sum_{k=1} \rho_k \exp\left[\frac{-(r-\delta_k)^2}{2\sigma_k^2}\right] \tag{4}$$

with the peak position δ_k, amplitude ρ_k, and width σ_k with k indicating *in, out, tail, inner, outer*, for each of the Gaussians, representing the headgroups of the two leaflets, the phospholipid tail region, and the inner and outer membrane decorations, respectively. The radius R is the distance defined from the center of the vesicle to δ_{tail}. Thus, we define $\varepsilon_k = \delta_k - R$ and fix $\varepsilon_{tail} = 0$.

The membrane thickness is characterized by the total thickness of the bilayer structure, α, defined by:

$$\alpha = \left[\varepsilon_{in} - \sigma_{in}\sqrt{2\pi}\right] - \left[\varepsilon_{out} - \sigma_{out}\sqrt{2\pi}\right] \tag{5}$$

SAXS data allow to measure the scattering contrast $\Delta\rho = \rho(r) - \rho_{solvent}$. This analysis provides the relative amplitudes of each bilayer and not the absolute amplitudes and the magnitude of $\rho_{solvent}$ was set equal to zero. The electronic density of the two peaks representing the headgroups were defined to be $\rho_{in,out} > 0$, and the peak that represents the lower electronic density of the bilayer, plausibly corresponding to the methyl groups of the phospholipids, was fixed to $\rho_{tail} = -1$ [24]. These two assumptions reduced the parameter dependency of the fitting procedure and also provided a means of comparing the models since all results were normalized to the central region of the bilayer. Notice that the Gaussians sum, representing the bilayer profile, and the Gaussian decorations, representing the protein, inter-penetrate to some extent. For a perfectly spherical, radially symmetric vesicle composed of n Gaussian shells, the form factor $<F(Q)^2> = F(Q)*F(Q)$ is obtained from the radially symmetric Fourier transform of Eq. (4). We used a normalized ensemble average over $F(Q)^2$ [19,22,23], resulting in

$$\langle F(Q)^2 \rangle = \varsigma \left\{ \frac{1}{Q^2} \sum_{k,k'} (R + \varepsilon_k)(R + \varepsilon'_k) p_k p'_k \sigma_k \sigma'_k \times \exp\left[Q^2 \left(\frac{\sigma_k^2 + \sigma_{k'}^2}{2} \right) \right] \cos\left[Q(\varepsilon_k - \varepsilon'_k) \right] \right\} \tag{6}$$

where $F(Q)^2$ is the form factor and ζ is a proportionality factor. Note that Eq. (6) is an approximation valid only in the region $0.1\ \text{nm}^{-1} > Q > 10\ \text{nm}^{-1}$, where intensity, arising from intra-bilayer features, dominates the scattering curve. The Eq. (6) was obtained from a normalized ensemble average over $F(Q)^2$, utilizing a Gaussian weight to describe the distribution of vesicles with radii R', average radius R and standard deviation $\sigma_{R'}$, as described in the Appendix A of ref. [19]. The data fitting procedure utilized the non-linear, least-squares Levenberg-Marquardt fitting (NLSF) algorithm to minimize χ^2, defined as:

$$\chi^2 = \sum_j \frac{\left(I(Q_j) - \hat{I}(Q_j)\right)^2}{\sigma(Q_j)} \tag{7}$$

where $\hat{I}(Q_j)$ is the fit value (model value) for Q_j, $I(Q_j)$ is the measured data value for the jth point and $\sigma(Q_j)$ is an estimate of the standard deviation for $I(Q_j)$. The electronic density profile of the lipid vesicle wall was obtained with high accuracy by fitting the model to the experimental data. This method allows also a good estimate of the average vesicle radius. Interaction of the membrane with other external components, such as the amyloid β-peptide induces a change in the bilayer structure that can be detected by SAXS.

2.4 Neutron spin echo

Created in the 70s, Neutron Spin Echo (NSE) (see e.g. Ref. [25]) is an extremely powerful technique for the study of the bilayer dynamics in membrane systems. Usually, in neutron scattering techniques, the scattering function $S(Q,\omega)$, as a function of the wave vector (Q) and the energy exchanged by the neutrons with the sample ($E = \hbar\omega$), is measured and then the intermediate scattering function $S(Q,t)$ can be obtained by time Fourier transformation. In contrast, NSE gives direct access to the intermediate scattering function $S(Q,t)$, that in turn, is the space Fourier transform of the Van Hove space-time correlation function $G(r,t)$, which reflects the structural and dynamical properties of the system. In other neutron scattering techniques, the determination of the energy exchanged by the neutrons with the samples requires a careful definition of the energy of both incoming and scattered neutron. As a consequence, a good energy resolution clearly implies a strong reduction in the number of detected neutrons. In NSE, the Larmor precession of the neutron magnetic moment in suitably designed magnetic fields is used to gain information directly on the energy difference between incident and scattered neutrons, thus allowing a much less stringent energy selection. After an initial time interval of dephasing for the spin of neutrons, a proper magnetic field reverts the spin orientations, so that, after the same time interval, a strong "echo" signal is observed, due to the restoration of the polarization. If, following the interaction with the sample, neutrons change their energy, and hence their velocity, each neutron spends a different time in the magnetic field region, this reducing the extent of polarization restoration. As a consequence, the measured scattered intensity lowers. The intermediate scattering function $S(Q,t)$ is obtained from the decay of the polarization.

In the interaction with membrane systems, neutrons may change their energy due to the elastic properties of the lipid bilayer, also influenced by energy dissipation at the membrane-water interface. Following the pioneering work of Zilman and Granek [26], the time decay of the intermediate scattering function can be modeled by a stretched exponential behavior:

$$\frac{S(Q,T)}{S(Q)} = e^{-(\Gamma t)^{2/3}} \tag{8}$$

where the relaxation rate Γ depends on the bending elasticity of the bilayer and is given by Refs. [27–31]:

$$\Gamma(Q) = 0.0069\gamma_k \sqrt{\frac{K_B T}{\kappa_c}} \frac{K_B T}{\eta} Q^3 \tag{9}$$

Here, κ_c is the bending modulus of the bilayer, η the viscosity of the solvent (typically D_2O). The factor γ_k originates from the averaging of the angle between the wavevector and the surface normal and approaches unity when $\kappa_c / K_B T \gg 1$, which is usually the case. It has been recently pointed out that one more factor must be considered in Eq. (8), to take into account the contribution of the diffusional translation of the whole nanoparticles [31–33]. This term changes Eq. (8) in

$$\frac{S(Q,T)}{S(Q)} = e^{-D_t Q^2 t} e^{-(\Gamma t)^{2/3}} \tag{10}$$

where D_t is the diffusion coefficient of the nanoparticle.

In a NSE experiment, the polarization decay is measured in a time scale from ~10^{-10} to ~10^{-7}, for several Q values (see Fig. 3), which can be properly determined by means of small angle neutron scattering measurements

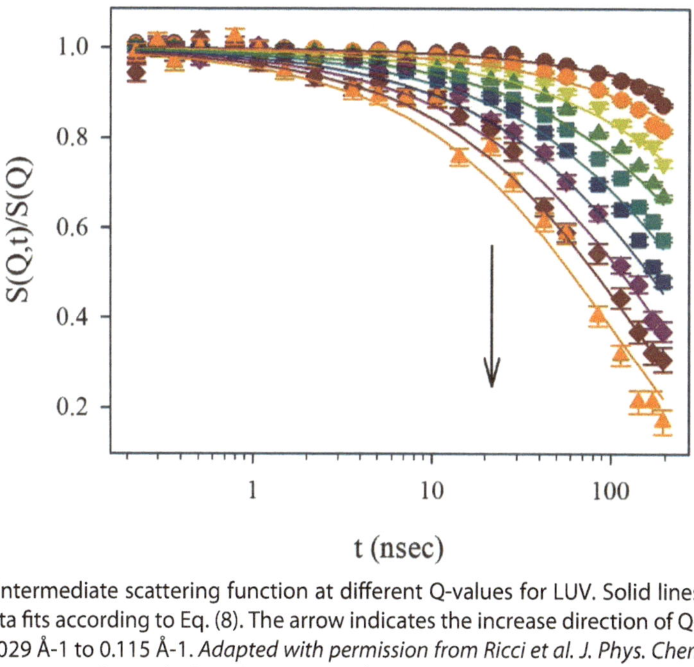

Fig. 3 Intermediate scattering function at different Q-values for LUV. Solid lines represent data fits according to Eq. (8). The arrow indicates the increase direction of Q-values, from 0.029 Å-1 to 0.115 Å-1. *Adapted with permission from Ricci et al. J. Phys. Chem. B 123 (2019) 631–638. Copyright {2019} American Chemical Society .*

on analogous samples. From the analysis of the various polarization decays it should be possible to obtain both the diffusion coefficient D_t and the bending modulus D_t of the lipid bilayer. In practice, however, data are often not sufficiently resolved for a simultaneous determination of both variables. In such cases, the value of the diffusion coefficient D_t can be obtained from dynamic light scattering experiments (see above) and used for a suitable analysis of NSE data. However, by comparing results from samples with different properties, it appears evident that the analysis with or without the overall diffusional term lead to qualitatively analogous conclusions, differing only in the actual values of the bending modulus. In these cases, analysis without diffusional term can be preferred for sake of simplicity.

3. Role of lipid composition in Aβ-membrane interaction

In the amyloid hypothesis, according to which the cause of the disease is the Aβ amyloid aggregation and deposition in the extracellular space (see e.g., Refs. [35,36]), it is believed that membrane can act as scaffold and play an important role in AD. Several decades ago, diverse studies brought to the intuition of the existence of functional regions in membranes, called lipid rafts, presenting an ordered phase and lower fluidity [37,38]. These membrane patches are rich in cholesterol (Chol), gangliosides and other sphingolipids, such as sphingomyelin (SM), and strongly differ in their lipid composition from the surrounding non-raft domains [39]. This special membrane environment fulfills the conditions for intracellular protein trafficking, protein-lipid and protein-protein interactions as well as transmembrane signaling [40,41]. In particular, the generation of Aβ has been shown to mainly take place in lipid rafts [42—45], thus indicating that a modulation of the membrane lipid composition might influence Aβ-generation. Moreover, the link between lipids and Aβ is also based on *post mortem* lipidomic studies in AD brains [46—48].

In this wide and complex contest, many studies address the effect of the membrane lipid composition in Aβ-peptide interaction, also considering the role that the lipid phase can play. The membrane phase is determined by the phospholipid composition, as well as the presence of sphingolipids and cholesterol. From the phase diagram of lipid ternary mixture of POPC/Chol/SM, studied on liposomes, it is evident that different phase regimes (L_o, a liquid ordered phase, L_d, a liquid disordered phase and S_o, a solid-like ordered phase) as well as their coexistence can be explored, depending on the ratio of the three components in the matrix [49]. The L_d phase is characterized by fast translational diffusion and low acyl chain order, the L_o phase is characterized by fast diffusion and high chain order, while the S_o phase is characterized by slow diffusion and high chain order [50].

The role of the bilayer charge has been reported, suggesting that electrostatic interactions play a role in the lipid surface-Aβ interaction [51–53]. The role of SM and SM with Cholesterol, some of the main components in lipid rafts, has also been studied. SM is a promoter of phase separation and coexistence of L_o/L_d phases, due to its ability to form hydrogen bonding with cholesterol or other components [54]. In the presence of Aβ, concentrated in the micromolar range, the aggregation is favored by SM especially in SM/Chol containing lipid systems in the ordered phase, as seen in supported lipid bilayers (SLB) [55]. Cholesterol *in se* is believed to be fundamental in the interaction with Aβ. In fact, the brain is the organ with the highest amount of cholesterol (constituting more than 30% of the lipid brain content) [56]. Seghezza et al. [57] showed that, in the presence of SM, a 5% cholesterol molar concentration leads to Aβ-membrane interaction with SLB. With higher cholesterol concentration (20%), a depletion in the interaction was observed, suggesting a preferential interaction of Aβ with the L_d lipid phase [57]. Interaction of POPC/POPS SLB without cholesterol (in the L_d phase) with micromolar concentrations of Aβ, both in the monomeric and in the early aggregation state has been reported, while no effect of the interaction was detected with Aβ mature fibrils [58]. Amaro et al. [59], studying Aβ-GUV interaction at nM Aβ concentrations by fluorescence correlation spectroscopy, address the interaction of the lipid matrix with Aβ monomer. From this starting point, they report that simple phospholipid membranes (made of different PC chains), without and with Chol do not prompt oligomeric formation on the membrane [59]. Further, they use disordered phase lipid systems, by mixing in the bilayer physiologically relevant components (DOPC, Chol, SM and GM1). They followed the unconventional idea that rafts are not rigid and ordered but highly dynamic nanoscopic regions [60]. They show that Aβ spontaneously forms oligomers on DOPC/Chol/SM GUVs as well as in DOPC/SM bilayers, differently than pure DOPC, and DOPC/Chol membranes, clearly pointing out the role of the SM in the oligomerization. This result is confirmed also by the inhibition of the oligomeric formation following substitution of SM with DSPC in the GUV matrix [59].

4. Role of ganglioside GM1 in Aβ-membrane interaction

Gangliosides are glycosphingolipids abundant in neurons, involved in many neurobiological pathways and in particular in Aβ production and oligomerization. Yanagisawa et al. in 1995 found in autopsied brain membranes the ganglioside GM1 complexed with Aβ (GAβ) [12]. It has also been reported that an antibody specific for N-terminal Aβ42 (BAN052) recognizes Aβ bound to GM1 in AD cerebral cortices, being the immunoreactivity correlated with cholera toxin reactivity, which simultaneously recognizes GM1 [61]. Ganglioside-bound Aβ exhibited seeding abilities for amyloid formation [62]. The interaction GM1-Aβ has been studied extensively in vitro, on different conditions and systems, all the studies demonstrating a specific interaction [63,64].

However, in spite of the number of studies, the involvement of GM1 in AD appears controversial. As above reported, it has been suggested that GM1 could act as a seed in Aβ amyloid formation [64,65] as well as, from cell line studies, that it could have a neuroprotective action [66]. This apparent contradiction can be overcome by considering the tendency of GM1 to form clusters, in a concentration dependent manner. Other colipids can also play a role in the Aβ-GM1 interaction, together with the range of Aβ concentration [67]. In fact, Matsuzaki and coworkers suggested different regimes in the Aβ-GM1 clusters interaction in the presence of sphingomyelin and cholesterol, depending on the molar ratio $r = \text{Lipids}/\text{A}\beta$ [68]. The interaction promotes the formation of alpha helical Aβ conformation for $r>80$, oligomerization for $20 < r<80$ and amyloid fibris on the membrane for $r<20$. In this case as well as in many other reports the GM1 molar concentration was however higher than 20%, resulting in a complete covering of the surface by the glycohead-groups. Levels of GM1 in brain membranes are 2–4%, and it seems that total ganglioside concentration decreases by aging, suggesting that GM1 could have a functional role of protection. Looking at data taken from model membranes, cultured cells and in vivo studies, different GM1 organization can influence differently Aβ oligomerization. Cebecauer et al. [67] give an exhaustive summary of the current knowledge, concluding that both effects, catalytic or inhibitory, can realize, depending on the GM1 clustering. However, the function/disfunction of GM1 in vivo is poorly understood. Consistently, Amaro et al. [59] reported that in a physiological mimetic regime (GM1 4%, Aβ nM) with a liquid disordered matrix containing also sphingomyelin, the presence of

GM1 inhibited the oligomerization, which on the contrary was observed without the ganglioside, thus suggesting a specific interaction of GM1 with Aβ monomers. A study performed by some of us uses SAXS to determine the structural changes in the bilayer due to the interaction of Aβ40 monomers with POPC/POPS/Chol unilamelar vesicles in *Ld* phase, without and with 2% ganglioside embedded in the matrix [15]. A suitable sum of Gaussians was used to model the electronic density profile, whose Fourier transform was obtained to fit the experimental data (see Fig. 2 above). Fig. 4 reports the SAXS spectra of LUV without and with GM1 alone and in the presence of Aβ (A) and the corresponding electronic density

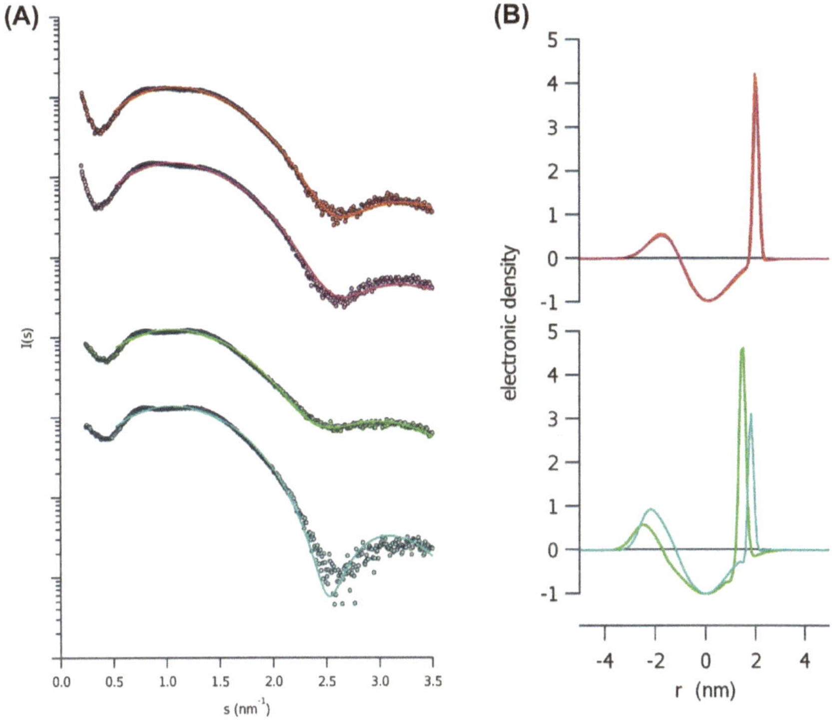

Fig. 4 I(s) = *I*(*Q*), SAXS spectra of PC:PS:Chol and PC:PS:Chol:GM1 LUVs before (red and green, respectively) and after mixing with Aβ. Lipid concentration was 15 mM and, when present, Aβ concentration was 50 μM. Continuous lines represent the best fitting of data, according to the multi-Gaussians model (A). Electronic density profiles for the sample PC:PS:Chol and PC:PS:Chol:GM1, before (red and green, respectively) and after mixing with Aβ, calculated from the experimental SAXS data according to the multi-Gaussians model (B). *Reprinted with permission from Nicastro et al. Biophys Chem 208 (2016) 9–16. Copyright 2016 Elsevier. M.C. Nicastro, D. Spigolon, F. Librizzi, O. Moran, M.G. Ortore, D. Bulone, P.L. San Biagio, R. Carrotta, Amyloid β-peptide insertion in liposomes containing GM1-cholesterol domains, Biophys. Chem. 208 (2016) 9–16.*

Table 1 SAXS Analysis: Vesicle parameters, *d* and *R*, obtained from data fits by the Gaussian model.

	PC:PS:Chol	PC:PS: Chol + Aβ	PC:PS:Chol: GM1	PC:PS:Chol: GM1+Aβ
correlation coefficient	0.9951	0.9928	0.9933	0.9935
bilayer thickness (nm)	3.59 ± 0.11	3.59 ± 0.06	3.91 ± 0.08	4.10 ± 0.11
vesicle radius (nm)	76.8 ± 1.7	80.2 ± 1.7	81.9 ± 1.8	86.3 ± 1.7

(B). The ratio LUV:Aβ was 300 and the ratio GM1:Aβ was 6:1. The effect of GM1 on the interaction is evident; in fact, an increase of the bilayer width of about 5% is detected only in the GM1 containing matrix (see Table 1).

Over time the system did not show any aggregation, thus confirming that the interaction does not prompt oligomerization and further aggregation. This is in agreement with the fact that in the absence of sphingomyelin the interaction concerns monomers and GM1 molecules, whose concentration does not favor large cluster formation [15]. Once established the preferential interaction with the ganglioside, the same matrix, organized in ~100 nm extruded LUV, was used to study the bilayer dynamical properties by means of NSE, which allows to measure the elastic bending constant of the lipid bilayer. The effects of Aβ in monomeric and oligomeric conditions were studied. In both cases, the presence of Aβ species lead to an increase in the rigidity of the membrane, as reflected by the elastic bending modulus of the bilayer (see Fig. 6B below) [34]. The effect of the presence of the chaperonin Hsp60 on the GM1-Aβ interaction has also been studied, as reported in the following.

5. Chaperonin HSP60 sequestration

Recent studies have been focused on the role of molecular chaperones in suppressing the consequences of protein misfolding and aggregation in neurodegenerative diseases [69–71]. Thanks to their structural and functional versatility, these biomolecules are able to play diverse roles from "folding" of non-native proteins to "holding" and/or degradation of harmful aggregates.

In particular, the class of molecular chaperones, presenting a characteristic double-ring oligomeric structure organized as containers for the folding of other proteins, revealed therapeutic potentiality in amyloid diseases like AD. In fact, it has been found by NMR measurements that the bacterial chaperonin GroEL suppresses $A\beta_{40}$ amyloid formation by interacting with its two hydrophobic segments Leu17-Ala21 and Ala30-Val36 that are key residues in fibrils formation [72]. Moreover, it has been recently shown that GroEL abrogates $A\beta_{42}$ neurotoxicity by binding to the peptide at three anchor-binding regions (residues 16–21, 31–34, and 40–41) [73]. The latter region is responsible of the higher neurotoxicity of $A\beta_{42}$ with respect to $A\beta_{40}$ [74].

Also Hsp60, the GroEL human homolog, seems to have a role in AD, being involved in APP translocation [75] and exerting a protective role, that results in the reduction of Aβ peptide dysfunctional synaptic binding [76]. Our previous studies showed that Hsp60 is able to inhibit both $A\beta_{40}$ and $A\beta_{42}$ amyloid aggregation. The inhibition mechanism was described as a specific recognition and sequestration of those peptide species, acting as seeds in the amyloid formation process, resulting in a dose dependent aggregation delay [77,78]. This hypothesis was corroborated by investigating the influence of the chaperonin on the Aβ-membrane interactions in a GM1 containing lipid matrix. By means of neutron spin echo measurements, both the molecular basis of Aβ toxicity and the chaperonin potential inhibition mechanisms were addressed. Aβ40 aggregation kinetics was studied, and initial and intermediate states were characterized by conformational analysis, both by Thioflavin T (ThT) fluorescence and circular dichroism, as reported in Fig. 5.

The bending modulus of LUV alone and in the presence of Aβ was obtained by NSE data analysis. Aβ was mixed to the LUV solution in monomeric state (freshly dissolved) and in oligomeric state (after incubation for 65 min at $T = 37\ ^\circ C$ under stirring).

The bending constant values were compared to the ones found after mixing the membrane solution with the corresponding Aβ/Hsp60 species, obtained under the same conditions as the Aβ species alone, i.e., by incubating Aβ with Hsp60, for $t = 0$ and $t = 65$ min at $T = 37\ ^\circ C$ and under stirring (Fig. 6B). Comparison of the ThT signal, coming from staining $A\beta_{40}$ species after 65 min of incubation, without and with Hsp60, is reported in Fig. 6A. In the presence of the chaperonin, ThT signal clearly shows that no aggregates are formed yet, as demonstrated in the kinetics studies reported in Mangione et al. 2016 [77].

The increase of the bilayer bending modulus observed for the vesicles when mixed with Aβ monomers as well as Aβ oligomers, confirmed that both species interact with the LUV. Very interestingly, in presence of the species Aβ/Hsp60 the membrane bending modulus was undistinguishable

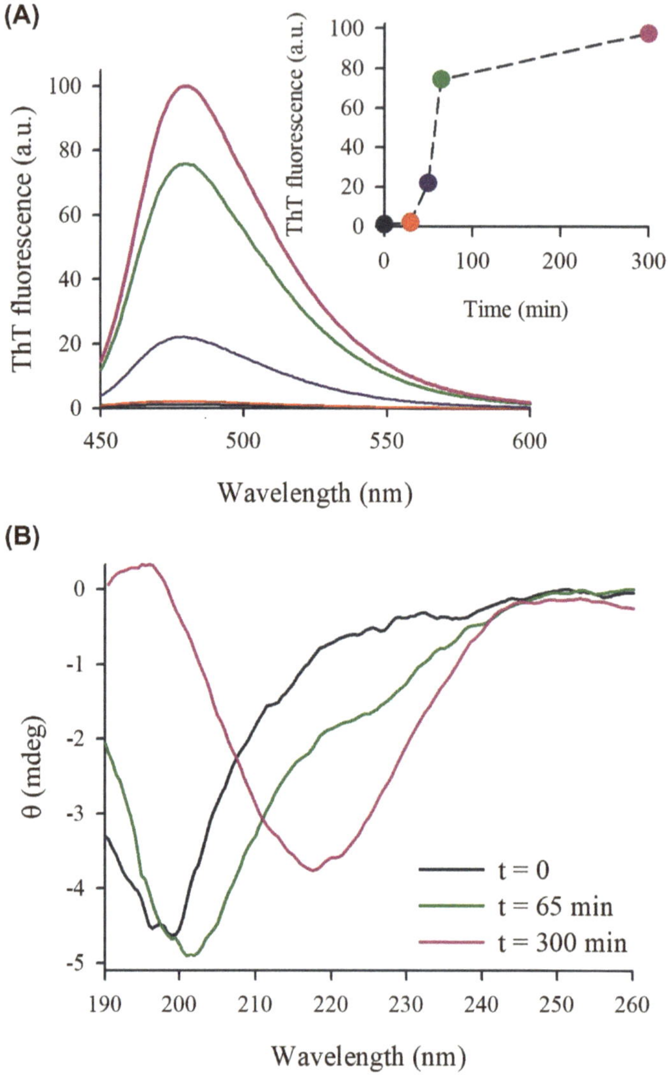

Fig. 5 Aggregation of the $A\beta_{1\text{-}40}$ peptides. A: ThT emission of 45 μM $A\beta_{1\text{-}40}$ at different times at 37 °C and $\lambda_{ex} = 450$ nm. Samples were measured at 0 min (black), 30 min (red), 50 min (blue), 65 min (green) and 300 min (violet). In the inset, the emission maximum at 476 nm is shown as a function of time. B: CD spectra of 45 μM $A\beta_{1\text{-}40}$ at 0 min (black), 65 min (green), so-defined oligomers in the text, and 300 min (violet). *Reprinted with permission from Ricci et al. J. Phys. Chem. B 123 (2019) 631–638. Copyright {2019} American Chemical Society .*

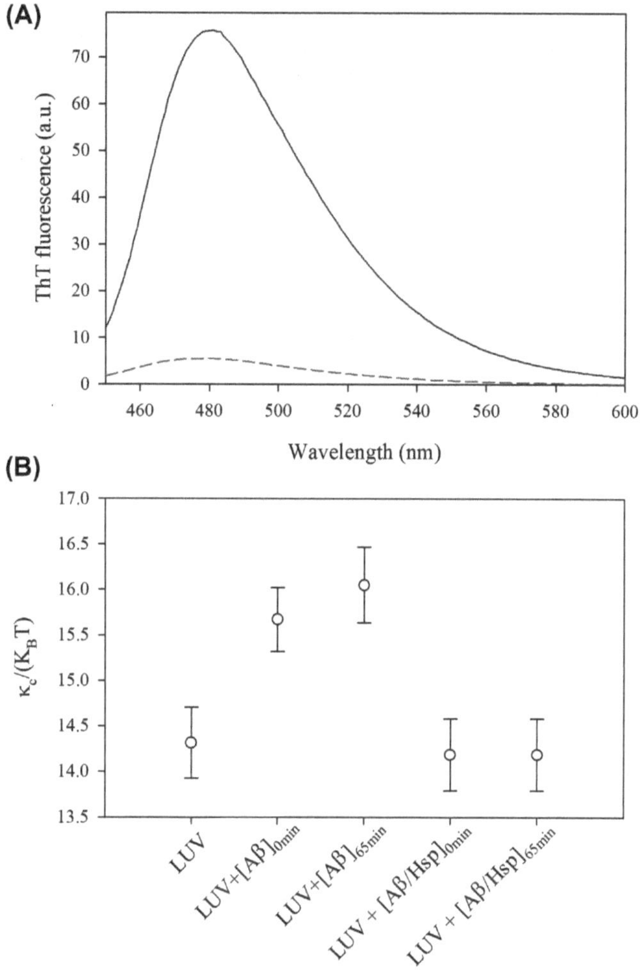

Fig. 6 A: ThT emission with $\lambda_{ex} = 450$ nm after mixing the dye with a 45 μM $A\beta_{1\text{-}40}$ solution incubated under stirring at 37 °C with 1.8 μM Hsp60 and collected after 65 min from the beginning of the aggregation process (red dashed line) and with a 45 μM $A\beta_{1\text{-}40}$ solution after 65 min of incubation under stirring at 37 °C (blue line). B: Comparison of the bending modulus from NSE data analysis for all the samples, as in the label. *Reprinted with permission from Ricci et al. J. Phys. Chem. B 123 (2019) 631–638. Copyright {2019} American Chemical Society* .

from the one obtained for unaffected LUV systems, thus suggesting that the GM1-Aβ interaction in our conditions interests Aβ species that are successfully recognized and sequestered by Hsp60. In the case of Aβ oligomeric species the result could be expected, since Hsp60 acts by sequestering those reactive species having a tendency to form intermediate oligomers and onset the aggregation cascade. Formation of these species was therefore inhibited in the presence of Hsp60 [77,78]. Instead, the result obtained in the case of freshly dissolved peptide in the presence of Hsp60 is most worthy of attention. In fact, it is known that in this concentration regime GM1 may interact with monomeric species [67]; therefore, reactive species may form directly on the surface of the LUV, due to GM1 presence [12], this process being hindered by the presence of Hsp60, even at substoichiometric concentration (1:25, Hsp60:Aβ). Hsp60 works as a chaperone, intervening in the Aβ-membrane interaction. Furthermore, in general, Hsp60 could also detect unmeasurable amounts of Aβ species, able to interact with the membrane and to onset aggregation. The understanding of the nature and origin of these species could represent an important target to setup a chaperone-based therapeutic strategy.

6. Conclusion

Despite the huge number of studies performed to investigate the neurotoxicity of Aβ species in membrane environment, no coherent conclusion can be drawn yet. This arises from the complexity of the in vivo problem. The understanding of why and how the eventual interaction takes place is still vague. Moreover, beside the difficulty to mimic the diverse and complex biological membranes and all the components that are playing a role, also the heterogeneity and instability of Aβ species makes the modeling difficult. However, although all these aspects prevent the full understanding of the AD etiology and of the membrane involvement in it, results collected in the last three decades have highlighted some relevant aspects and components. It has become evident that simple models are useful for basic understanding but that AD biology involves a bidirectional link between lipid membrane components and Aβ production and aggregation.

References

[1] G.S. Bloom, Amyloid-beta and tau: the trigger and bullet in Alzheimer disease pathogenesis, JAMA Neurol. 71 (2014) 505–508.

[2] W. Gulisano, D. Maugeri, M.A. Baltrons, M. Fa, A. Amato, A. Palmeri, L. D'Adamio, C. Grassi, D.P. Devanand, L.S. Honig, D. Puzzo, O. Arancio, Role of amyloid-beta and tau proteins in Alzheimer's disease: confuting the amyloid cascade, J. Alzheimers. Dis. 68 (2019) 415.

[3] B. Penke, F. Bogar, G. Paragi, J. Gera, L. Fulop, Key peptides and proteins in Alzheimer's disease, Curr. Protein Pept. Sci. 20 (2019) 577–599.

[4] A.S. DeToma, S. Salamekh, A. Ramamoorthy, M.H. Lim, Misfolded proteins in Alzheimer's disease and type II diabetes, Chem. Soc. Rev. 41 (2012) 608–621.

[5] P.J. Knowles, M. Vendruscolo, C.M. Dobson, The amyloid state and its association with protein misfolding diseases, Nat. Rev. Mol. Cell Biol. 15 (2014) 384–396.
[6] F. Chiti, C.M. Dobson, Protein misfolding, amyloid formation, and human disease: a summary of progress over the last decade, Annu. Rev. Biochem. 86 (2017) 27–68.
[7] R.A. Nixon, Amyloid precursor protein and endosomal-lysosomal dysfunction in Alzheimer's disease: inseparable partners in a multifactorial disease, FASEB J. 31 (2017) 2729–2743.
[8] H.A. Bustamante, A.E. Gonzalez, C. Cerda-Troncoso, R. Shaughnessy, C. Otth, A. Soza, P. V Burgos, Interplay between the autophagy-lysosomal pathway and the ubiquitin-proteasome system: a target for therapeutic development in Alzheimer's disease, Front. Cell. Neurosci. 12 (2018) 126.
[9] V.W. Chow, M.P. Mattson, P.C. Wong, M. Gleichmann, An overview of APP processing enzymes and products, NeuroMolecular Med. 12 (2010) 1–12.
[10] O. Wirths, G. Multhaup, C. Czech, V. Blanchard, S. Moussaoui, G. Tremp, L. Pradier, K. Beyreuther, T.A. Bayer, Intraneuronal Abeta accumulation precedes plaque formation in beta-amyloid precursor protein and presenilin-1 double-transgenic mice, Neurosci. Lett. 306 (2001) 116–120.
[11] F.M. LaFerla, K.N. Green, S. Oddo, Intracellular amyloid-beta in Alzheimer's disease, Nat. Rev. Neurosci. 8 (2007) 499–509.
[12] K. Yanagisawa, A. Odaka, N. Suzuki, Y. Ihara, GM1 ganglioside-bound amyloid beta-protein (A beta): a possible form of preamyloid in Alzheimer's disease, Nat. Med. 1 (1995) 1062–1066.
[13] D.D. Lasic, The mechanism of vesicle formation, Biochem. J. 256 (1988) 1–11.
[14] A. Dickey, R. Faller, Examining the contributions of lipid shape and headgroup charge on bilayer behavior, Biophys. J. 95 (2008) 2636–2646.
[15] M.C. Nicastro, D. Spigolon, F. Librizzi, O. Moran, M.G. Ortore, D. Bulone, P.L. San Biagio, R. Carrotta, Amyloid β-peptide insertion in liposomes containing GM1-cholesterol domains, Biophys. Chem. 208 (2016) 9–16.
[16] P. Stepanek, Data analysis in dynamic light scattering, in: W. Brown (Ed.), Dyn. Light Scatt. Method Some Appl., Clarendon Press, Oxford, UK, 1993, p. 177.
[17] A. Guinier, X-ray Diffraction in Crystals, Imperfect Crystals, and Amorphous Bodies, Courier Dover Publications, San Francisco and London, 1963.
[18] R. Carrotta, J. Barthès, A. Longo, V. Martorana, G. Portale, P.L. San Biagio, Eur. Biophys. J. 36 (2007) 701–709.
[19] M.R. Brzustowicz, A.T. Brunger, X-ray scattering from unilamellar lipid vesicles, J. Appl. Crystallogr. 38 (2005) 126–131.
[20] S. Castorph, L. Arleth, M. Sztucki, U. Vainio, S.K. Ghosh, M. Holt, J. R, T. Salditt, J. Phys. Conf. Ser. 247 (2010) 012015.
[21] S. Castorph, D. Riedel, L. Arleth, M. Sztucki, R. Jahn, M. Holt, T. Salditt, Structure parameters of synaptic vesicles quantified by small-angle x-ray scattering, Biophys. J. 98 (2010) 1200–1208.
[22] D. Baroni, O. Zegarra-Moran, A. Svensson, O. Moran, Direct interaction of a CFTR potentiator and a CFTR corrector with phospholipid bilayers, Eur. Biophys. J. 43 (2014) 341–346.
[23] D. Baroni, O. Zegarra-Moran, O. Moran, Functional and pharmacological induced structural changes of the cystic fibrosis transmembrane conductance regulator in the membrane solved using SAXS, Cell. Mol. Life Sci. 72 (2015) 1363–1375.
[24] G. Pabst, M. Rappolt, H. Amenitsch, P. Laggner, Structural information from multilamellar liposomes at full hydration: full q-range fitting with high quality x-ray data, Phys. Rev. E: Stat. Phys., Plasmas, Fluids, Relat. Interdiscip. Top. 62 (2000) 4000–4009.

[25] F. Mezei, The principles of neutron spin echo, in: Neutron Spin Echo, Lect. Notes Physics, vol. 128, 1980, pp. 1–26.
[26] G. Zilman, Undulations and dynamic structure factor of membranes, Phys. Rev. Lett. 77 (1996) 4788–4791.
[27] Z. Yi, M. Nagao, D.P. Bossev, Bending elasticity of saturated and monounsaturated phospholipid membranes studied by the neutron spin echo technique, J. Phys. Condens. Matter 21 (2009) 155104.
[28] M.C. Watson, F.L.H. Brown, Interpreting membrane scattering experiments at the mesoscale: the contribution of dissipation within the bilayer, Biophys. J. 98 (2010) L9–L11.
[29] M. Nagao, E.G. Kelley, R. Ashkar, R. Bradbury, P.D. Butler, Probing elastic and viscous properties of phospholipid bilayers using neutron spin echo spectroscopy, J. Phys. Chem. Lett. 8 (2017) 4679–4684.
[30] S. Chakraborty, A. Abbasi, G.D. Bothun, M. Nagao, C.L. Kitchens, Phospholipid bilayer softening due to hydrophobic gold nanoparticle inclusions, Langmuir 34 (2018) 13416–13425.
[31] S. Gupta, J.U. De Mel, R.M. Perera, P. Zolnierczuk, M. Bleuel, A. Faraone, G.J. Schneider, Dynamics of phospholipid membranes beyond thermal undulations, J. Phys. Chem. Lett. 9 (2018) 2956–2960.
[32] M. Mell, L.H. Moleiro, Y. Hertle, P. Fouquet, R. Schweins, I. Lopez-Montero, T. Hellweg, F. Monroy, Bending stiffness of biological membranes: what can be measured by neutron spin echo? Eur. Phys. J. E. Soft Matter. 36 (2013) 75.
[33] I. Hoffmann, R. Michel, M. Sharp, O. Holderer, M.-S. Appavou, F. Polzer, B. Farago, M. Gradzielski, Softening of phospholipid membranes by the adhesion of silica nanoparticles–as seen by neutron spin-echo (NSE), Nanoscale 6 (2014) 6945–6952.
[34] C. Ricci, M. Maccarini, P. Falus, F. Librizzi, M.R. Mangione, O. Moran, M.G. Ortore, R. Schweins, S. Vilasi, R. Carrotta, Amyloid beta-peptide interaction with membranes: can chaperones change the fate? J. Phys. Chem. B 123 (2019) 631–638.
[35] J. Hardy, D.J. Selkoe, The amyloid hypothesis of Alzheimer's disease: progress and problems on the road to therapeutics, Science 297 (2002) 353–356.
[36] D.J. Selkoe, J. Hardy, The amyloid hypothesis of Alzheimer's disease at 25 years, EMBO Mol. Med. 8 (2016) 595–608.
[37] J.L. Ranck, L. Mateu, D.M. Sadler, A. Tardieu, T. Gulik-Krzywicki, V. Luzzati, Order-disorder conformational transitions of the hydrocarbon chains of lipids, J. Mol. Biol. 85 (1974) 249–277.
[38] R. Schroeder, E. London, D. Brown, Interactions between saturated acyl chains confer detergent resistance on lipids and glycosylphosphatidylinositol (GPI)-anchored proteins: GPI-anchored proteins in liposomes and cells show similar behavior, Proc. Natl. Acad. Sci. U. S. A. 91 (1994) 12130–12134.
[39] E. Posse de Chaves, S. Sipione, Sphingolipids and gangliosides of the nervous system in membrane function and dysfunction, FEBS Lett. 584 (2010) 1748–1759.
[40] D.A. Brown, J.K. Rose, Sorting of GPI-anchored proteins to glycolipid-enriched membrane subdomains during transport to the apical cell surface, Cell 68 (1992) 533–544.
[41] D. Lingwood, K. Simons, Lipid rafts as a membrane-organizing principle, Science 327 (2010) 46–50.
[42] S.J. Lee, U. Liyanage, P.E. Bickel, W. Xia, P.T.J. Lansbury, K.S. Kosik, A detergent-insoluble membrane compartment contains A beta in vivo, Nat. Med. 4 (1998) 730–734.
[43] D.R. Riddell, G. Christie, I. Hussain, C. Dingwall, Compartmentalization of beta-secretase (Asp2) into low-buoyant density, noncaveolar lipid rafts, Curr. Biol. 11 (2001) 1288–1293.

[44] R. Ehehalt, P. Keller, C. Haass, C. Thiele, K. Simons, Amyloidogenic processing of the Alzheimer beta-amyloid precursor protein depends on lipid rafts, J. Cell Biol. 160 (2003) 113–123.
[45] K.S. Vetrivel, H. Cheng, W. Lin, T. Sakurai, T. Li, N. Nukina, P.C. Wong, H. Xu, G. Thinakaran, Association of gamma-secretase with lipid rafts in post-Golgi and endosome membranes, J. Biol. Chem. 279 (2004) 44945–44954.
[46] M. Diaz, N. Fabelo, I. Ferrer, R. Marin, "Lipid raft aging" in the human frontal cortex during nonpathological aging: gender influences and potential implications in Alzheimer's disease, Neurobiol. Aging 67 (2018) 42–52.
[47] M.O.W. Grimm, S. Grosgen, M. Riemenschneider, H. Tanila, H.S. Grimm, T. Hartmann, From brain to food: analysis of phosphatidylcholins, lyso-phosphatidylcholins and phosphatidylcholin-plasmalogens derivates in Alzheimer's disease human post mortem brains and mice model via mass spectrometry, J. Chromatogr. A. 1218 (2011) 7713–7722.
[48] X. Han, Multi-dimensional mass spectrometry-based shotgun lipidomics and the altered lipids at the mild cognitive impairment stage of Alzheimer's disease, Biochim. Biophys. Acta 1801 (2010) 774–783.
[49] R.F.M. De Almeida, A. Fedorov, M. Prieto, Sphingomyelin/phosphatidylcholine/cholesterol phase diagram: boundaries and composition of lipid rafts, Biophys. J. 85 (2003) 2406–2416.
[50] D.G. Ackerman, G.W. Feigenson, Lipid bilayers: clusters, domains and phases, Essays Biochem. 57 (2015) 33–42.
[51] J.M. Alarcon, J.A. Brito, T. Hermosilla, I. Atwater, D. Mears, E. Rojas, Ion channel formation by Alzheimer's disease amyloid beta-peptide (Abeta40) in unilamellar liposomes is determined by anionic phospholipids, Peptides 27 (2006) 95–104.
[52] E. Hellstrand, E. Sparr, S. Linse, Retardation of Abeta fibril formation by phospholipid vesicles depends on membrane phase behavior, Biophys. J. 98 (2010) 2206–2214.
[53] T.L. Williams, L.C. Serpell, Membrane and surface interactions of Alzheimer's Abeta peptide–insights into the mechanism of cytotoxicity, FEBS J. 278 (2011) 3905–3917.
[54] J.P. Slotte, The importance of hydrogen bonding in sphingomyelin's membrane interactions with co-lipids, Biochim. Biophys. Acta 1858 (2016) 304–310.
[55] S. Devanathan, Z. Salamon, G. Lindblom, G. Gröbner, G. Tollin, Effects of sphingomyelin, cholesterol and zinc ions on the binding, insertion and aggregation of the amyloid Aβ1-40 peptide in solid-supported lipid bilayers, FEBS J. 273 (2006) 1389–1402.
[56] C. Di Scala, H. Chahinian, N. Yahi, N. Garmy, J. Fantini, Interaction of Alzheimer's beta-amyloid peptides with cholesterol: mechanistic insights into amyloid pore formation, Biochemistry 53 (2014) 4489–4502.
[57] S. Seghezza, A. Diaspro, C. Canale, S. Dante, Cholesterol drives abeta(1-42) interaction with lipid rafts in model membranes, Langmuir 30 (2014) 13934–13941.
[58] C. Canale, S. Seghezza, S. Vilasi, R. Carrotta, D. Bulone, A. Diaspro, P.L. San Biagio, S. Dante, Different effects of Alzheimer's peptide Abeta(1-40) oligomers and fibrils on supported lipid membranes, Biophys. Chem. 182 (2013) 23–29.
[59] M. Amaro, R. Šachl, G. Aydogan, I.I. Mikhalyov, R. Vácha, M. Hof, GM1Ganglioside inhibits β-amyloid oligomerization induced by sphingomyelin, Angew. Chem. Int. Ed. 55 (2016) 9411–9415.
[60] E. Sevcsik, G.J. Schutz, With or without rafts? Alternative views on cell membranes, Bioessays 38 (2016) 129–139.
[61] K. Yanagisawa, Y. Ihara, GM1 ganglioside-bound amyloid beta-protein in Alzheimer's disease brain, Neurobiol. Aging 19 (1998) S65–S67.
[62] A. Kakio, S. Nishimoto, K. Yanagisawa, Y. Kozutsumi, K. Matsuzaki, Interactions of amyloid beta-protein with various gangliosides in raft-like membranes: importance of

GM1 ganglioside-bound form as an endogenous seed for Alzheimer amyloid, Biochemistry 41 (2002) 7385–7390.
[63] T. Ariga, M.P. McDonald, R.K. Yu, Role of ganglioside metabolism in the pathogenesis of Alzheimer's disease—a review, J. Lipid Res. 49 (2008) 1157–1175.
[64] K. Yanagisawa, GM1 ganglioside and Alzheimer's disease, Glycoconj. J. 32 (2015) 87–91.
[65] K. Matsuzaki, How do membranes initiate Alzheimer's Disease? Formation of toxic amyloid fibrils by the amyloid beta-protein on ganglioside clusters, Acc. Chem. Res. 47 (2014) 2397–2404.
[66] F. Kreutz, R.L. Frozza, A.C. Breier, V.A. de Oliveira, A.P. Horn, L.F. Pettenuzzo, C.A. Netto, C.G. Salbego, V.M.T. Trindade, Amyloid-beta induced toxicity involves ganglioside expression and is sensitive to GM1 neuroprotective action, Neurochem. Int. 59 (2011) 648–655.
[67] M. Cebecauer, M. Hof, M. Amaro, Impact of GM1 on membrane-mediated aggregation/oligomerization of β-amyloid: unifying view, Biophys. J. 113 (2017) 1194–1199.
[68] K. Ikeda, T. Yamaguchi, S. Fukunaga, M. Hoshino, K. Matsuzaki, Mechanism of amyloid β-protein aggregation mediated by GM1 ganglioside clusters, Biochemistry 50 (2011) 6433–6440.
[69] S.N. Witt (Ed.), Protein Chaperones and Protection from Neurodegenerative Diseases, John Wiley & Sons, Inc., 2011.
[70] C.G. Evans, S. Wisen, J.E. Gestwicki, Heat shock proteins 70 and 90 inhibit early stages of amyloid beta-(1-42) aggregation in vitro, J. Biol. Chem. 281 (2006) 33182–33191.
[71] T. Hoshino, M.I. Mahmood, K. Mori, K. Matsuzaki, Binding and aggregation mechanism of amyloid β-peptides onto the GM1 ganglioside-containing lipid membrane, J. Phys. Chem. B 117 (2013) 8085–8094.
[72] M. Yagi-Utsumi, T. Kunihara, T. Nakamura, Y. Uekusa, K. Makabe, K. Kuwajima, K. Kato, NMR characterization of the interaction of GroEL with amyloid beta as a model ligand, FEBS Lett. 587 (2013) 1605–1609.
[73] M.A. Walti, J. Steiner, F. Meng, H.S. Chung, J.M. Louis, R. Ghirlando, V. Tugarinov, A. Nath, G.M. Clore, Probing the mechanism of inhibition of amyloid-beta(1-42)-induced neurotoxicity by the chaperonin GroEL, Proc. Natl. Acad. Sci. U. S. A. 115 (2018) E11924–E11932.
[74] J.T. Jarrett, E.P. Berger, P.T.J. Lansbury, The carboxy terminus of the beta amyloid protein is critical for the seeding of amyloid formation: implications for the pathogenesis of Alzheimer's disease, Biochemistry 32 (1993) 4693–4697.
[75] K.C. Walls, P. Coskun, J.L. Gallegos-Perez, N. Zadourian, K. Freude, S. Rasool, M. Blurton-Jones, K.N. Green, F.M. LaFerla, Swedish Alzheimer mutation induces mitochondrial dysfunction mediated by HSP60 mislocalization of amyloid precursor protein (APP) and beta-amyloid, J. Biol. Chem. 287 (2012) 30317–30327.
[76] C. Marino, B. Krishnan, F. Cappello, G. Taglialatela, Hsp60 protects against amyloid beta oligomer synaptic toxicity via modification of toxic oligomer conformation, ACS Chem. Neurosci. 10 (2019) 2858–2867.
[77] M.R. Mangione, S. Vilasi, C. Marino, F. Librizzi, C. Canale, D. Spigolon, F. Bucchieri, A. Fucarino, R. Passantino, F. Cappello, D. Bulone, P.L. San Biagio, Hsp60, amateur chaperone in amyloid-beta fibrillogenesis, Biochim. Biophys. Acta Gen. Subj. 1860 (2016).
[78] S. Vilasi, R. Carrotta, C. Ricci, G.C. Rappa, F. Librizzi, V. Martorana, M.G. Ortore, M.R. Mangione, Inhibition of abeta1-42 fibrillation by chaperonins: human Hsp60 is a stronger inhibitor than its bacterial homologue GroEL, ACS Chem. Neurosci. 10 (2019) 3565–3574.

CHAPTER TWO

Propolis flavonoids and terpenes, and their interactions with model lipid membranes: a review

Luka Šturm and Nataša Poklar Ulrih*
Department of Food Science and Technology, Biotechnical Faculty, University of Ljubljana, Ljubljana, Slovenia
*Corresponding author: E-mail: natasa.poklar@bf.uni-lj.si

Contents

Abstract

Propolis is a complex mixture of different biologically active compounds. Among these, phenols (especially flavonoids) and terpenes are the most important and abundant. As these compounds are gaining attention in development of suppressors of cancer and microbial disease, studies of their interactions with lipid membranes are becoming more and more relevant. In this article, the interactions of most of the important compounds from propolis (phenols: mostly flavonoids, and terpenes) with different kinds of model membranes are reviewed. Knowledge from these studies can be used for further development of new cancer suppressors and anti-microbial drugs, as well as for many more uses.

List of abbreviations

DODAB Dioctadecyldimethylammonium bromide
DOPC Dioleoylphosphatidylcholine

Advances in Biomembranes and Lipid Self-Assembly, Volume 32
ISSN 2451-9634
https://doi.org/10.1016/bs.abl.2020.04.003

DOPE Dioleoylphosphatidylethanolamine
DPPC Dipalmitoylphosphatidylcholine
DPPE Dipalmitoylphosphatidylethanolamine
DPPG Dipalmitoylphosphatidylglycerol
DPPS Dipalmitoylphosphatidylserine
POPC 1-palmitoyl-2-oleoylphosphatidylcholine
POPE 1-palmitoyl-2-oleoylphosphatidylethanolamine
POPS 1-palmitoyl-2-oleoylphosphatidylserine
SOPS 1-stearoyl-2-oleoylphosphatidylserine

1. Introduction

The cell membrane is one of the most important parts of the cell. Indeed, it defines the cell as such. Therefore, it is an integral part of bacterial, archeal, fungal, plant and animal cells. It is the boundary between the inner (intracellular) and outer (extracellular) environments, and it has an important role in cell functionality. It shields the inner cell components from the effects of the outside environment, while allowing the exchange of nutrients and other molecules between the cell and its surroundings. Cell membranes also have a key role in signaling and communicating between different cells [1−4]. Therefore, they have been extensively studied ever since their initial discovery. These studies have helped us to understand important cell phenomena like nerve function, drug resistance, uptake and release of different molecules, effects of environmental stress and its negation, and many others [4−8].

As the cellular shield, cell membranes also have a major role in prevention of cellular damage and death due to adverse effects of the environment or different environment molecules. On the other hand, cell membranes can be the main reason for cellular collapse following their malfunctioning, disruption and/or rupture. In this regard, they are a major target for antibiotics and other biologically important molecules, such as cancer therapeutics. As cell membranes also regulate the uptake and release of molecules of the cells (including drugs), knowledge of the interactions between specific molecules and cell membranes is of crucial importance [7,9−11].

Propolis is a chemically diverse substance that is known to include many biologically important compounds, including antioxidants, cancer inhibitors, anesthetics, antibacterial agents, immunostimulants, and others. As a substance, it is important in traditional medicine and it is also gaining importance in clinical medicine, as it has been shown to be an effective inhibitor of cancer cells and microorganisms [12]. Nevertheless, the very diverse compositions of propolis remain poorly characterized, and so it is difficult to determine to which molecules the specific properties linked to the actions of propolis can be attributed. Many completely new substances have been discovered in propolis, so their applications also remain unknown, and furthermore, for some of these compounds, their synergistic effects have been shown to be greater than the effects of the individual components alone [13–15].

Most drugs and biologically important molecules have specific target molecules within the cell. Drug–lipid membrane interactions thus occur at least at some point, which affects, sometimes severely, the pharmacokinetics (i.e., transport, distribution, accumulation) and efficacy of these molecules [16,17]. For these reasons, it is important to characterize specific propolis samples, to determine their overall impact and the effects of specific propolis compounds on cells and their membranes. This can significantly enhance their biological potential [15].

However, characterizing the properties of specific molecules and their specific effects in living cells is a particularly complex problem, so often artificial lipid membranes, or empty cells—"cell ghosts"—are used instead. Through a combination of different techniques, such as differential scanning calorimetry (DSC), electron paramagnetic resonance (EPR), and different fluorescence techniques, it is possible to gain better understanding of the mode of interaction between specific modulatory compounds and specific membrane lipids.

2. Propolis definition, its history and its characteristics

Propolis, or "bee glue", is a sticky, hydrophilic, gummy and resinous material that is produced by different species of bees, including honeybees (*Apis* spp.) and stingless bees (*Melipona* spp.). Its name derives from the Hellenistic Ancient Greek meaning "suburb/bee glue" or "defense of the city", depending on the interpretation. Bees use it to seal cracks in their hives, to smooth the internal walls, and to protect the hive against intruders. Propolis also acts as a natural antibiotic, to prevent bacterial, viral or fungal infections within the hive. Bees collect the resin from the cracks in the bark and leaf buds of different trees, including species of poplar, conifer (e.g., pine, cypress), birch, alder, willow, palm and chestnut, and even from trees like eucalyptus, acacia, *Clusia* spp. and *Baccharis dracunculifolia* (for Brazilian green propolis). Bees mix the collected resin with salivary enzymes and beeswax, and they use this partially digested material in their hives. The color of propolis depends on its primary plant source and age, and it can vary from yellow to green, and red to dark brown; there have even been reports of transparent propolis [18–23].

Propolis can be typified in several different ways. According to the bee group that produces it, two main types of propolis have been described. The first is the "classic" propolis that is produced by honeybees, and the second is "geopropolis", that is produced by stingless bees, which also add soil to their propolis mixtures [19,24]. According to the plant sources, propolis has been classified into seven types [25,26], which include poplar and Brazilian green propolis, which are the most well-known and used [27].

Propolis and its uses are as old as honey, as it was known to the ancient Egyptians, Persians and Romans. It has been used ever since as an important folk remedy [20]. Despite the progress in pharmacology, the list of preparations and uses of propolis even today remains enormous, mostly because of its antiseptic, bacteriostatic, antibacterial, antimycotic, antiviral, antiprotozoan, antioxidative, spasmolytic, choleric, astringent, anti-inflammatory, anesthetic, antitumor, immunostimulating, anticancer, cytostatic and hepatoprotective properties, among others [18,28].

The composition of propolis depends on many factors, with the main ones being its primary plant source, the season it is produced in, and the bee species that produces it. In general, it is composed of about 50% plant balsam and resin, 30% bees wax, 10% essential and aromatic oils, 5% pollen and 5% other organic and inorganic molecules. This is especially true for the poplar type of propolis [12,18]. By the beginning of the 21st century, Marcucci (1995) and Bankova et al. (2000) had described more than 300 constituents in propolis, and then just between 2000 and 2012, at least 241 more compounds were isolated from propolis for the first time, some of which were previously unknown to man [29]. By the present year, 2019, the number of compounds isolated from propolis has grown to a total of approximately 850 [112], and this continues to grow every year as new components are being discovered in propolis from unexplored regions and from new plant origins. However, each propolis sample contains approximately 80–100 different constituents [20]. Among these, there are phenolic acids and esters, many types of flavonoids and other phenolic molecules, terpenes, ketones, aromatic aldehydes and alcohols, proteins, fatty acids, waxy acids, amino acids, steroids, stilbenes, sugars, vitamins, minerals, and even enzymes [19,21,29–31] (Fig. 1).

The phenols are among the most important constituents of propolis, and on average they represent around 28% ($\pm$9%) of the whole mass of poplar

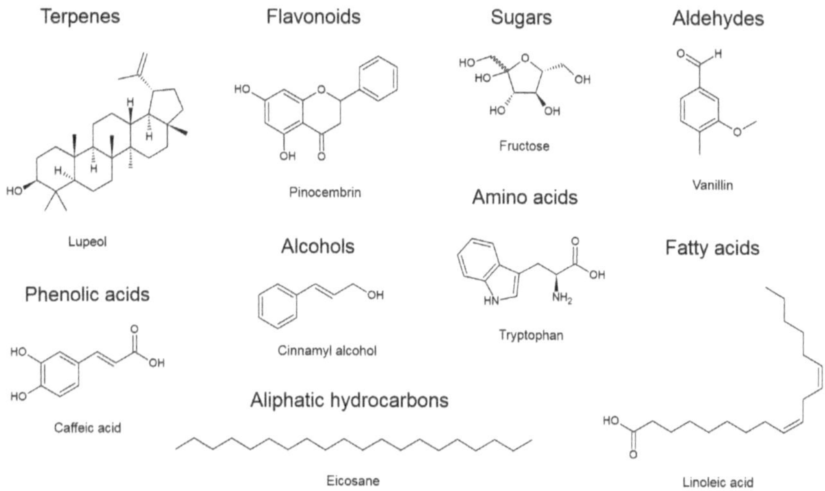

Fig. 1 Some of the more typical and abundant groups of compounds described for propolis.

propolis, among which 8% (±4%) are flavones/flavonols and 6% (±2%) are flavanones/dihydroflavonols. The phenols are also believed to be the most biologically active constituents of propolis, among which the flavonoids are especially active [27,29,32,33]. The other highly relevant biologically constituents of propolis are the terpenes. These have been mainly described for tropical propolis, which is rarer than poplar propolis, and they have also been isolated from Mediterranean propolis. Among these terpenes, the sesquiterpenes are biologically the most important [29,34,35].

The sheer amount of different biologically active components makes propolis one of the most biologically active substances known, and as propolis is gaining importance as an anticancer and antimicrobial agent, the need to study the interactions between its components and the cell functionality is also growing. As there are way too many compounds in propolis to review the full spectrum of their interactions with cell membranes, this review focusses only on some of the most abundant or biologically most important compounds in propolis. The specifics about these chosen compounds are given in Table 1. All of the topological polar surface-area data for these chosen compounds, as well as the number of the H-bonds they can form, were retrieved from the PubChem database [36,37]. These propolis constituents were chosen on the basis of research articles [13,27,33–35,38–43] and availability of studies regarding the interactions of these constituents with cell membranes.

Table 1 The most abundant and biologically important compounds in propolis, and their topological polar surface areas and numbers of H-bonds.

Compound	Type	Topological polar surface area (Å²)	H-bonds available To donate	H-bonds available To accept
Phenols and flavonoids				
Quercetin	Flavonol	127	5	7
Rutin	Flavonol glycoside	266	10	16
Galangin	Flavonol	87	3	4
Kaempferol	Flavonol	107	4	6
Chrysin	Flavone	66.8	2	4
Luteolin	Flavone	107	4	6
Apigenin	Flavone	87	3	5
Naringenin	Flavanone	87	3	5
Pinocembrin	Flavanone	87	2	4
Cinnamic acid	Hydroxycinnamic acid	37.3	1	2
ρ-Coumaric acid	Hydroxycinnamic acid	57.5	2	3
Ferulic acid	Hydroxycinnamic acid	66.8	2	4
Caffeic acid	Hydroxycinnamic acid	77.8	3	4
Caffeic acid phenethyl ester	Phenol	66.8	2	4
Artepillin C	Phenol	57.5	2	3
α-Tocopherol	Phenol/Vitamin	29.5	1	2
Eugenol	Phenol	29.5	1	2
Terpenes				
α-Amyrin	Triterpene	20.2	1	1
β-Amyrin	Triterpene	20.2	1	1
Isocupressic acid	Diterpene acid	57.5	2	3
Geraniol	Monoterpene	20.2	1	1
Limonene	Monoterpene	0	0	0
Lupeol	Triterpene	20.2	1	1

3. Artificial lipid membranes

The basic structural unit of biological membranes is the lipid bilayer. Lipid bilayers are sheet-like assemblies of countless amphiphilic lipid molecules where their structural arrangement is held together by hydrophobic interactions. These lipid bilayers form the boundary between the inner and outer cellular environments, as the plasma membrane, which effectively defines the cell; within the cell, lipid bilayers also define the intracellular organelles [10]. The plasma membrane contains lipids and proteins in different proportions (including glycolipids and glycoproteins, as 5%–10%); however, the lipid bilayer is considered as the basic building block of the cellular membrane [1].

Based on their chemical structures, membrane lipids can be divided in three main groups: phospholipids, ceramide-based sphingolipids, and cholesterol. Of these, the phospholipids are the main component of cell membranes [44]. They are molecules that contain a hydrophilic headgroup and hydrophobic acyl chains, which are linked together through an alcohol. Based on the hydrophilicity of their headgroups, the phospholipids are further divided into a number of subgroups. The most abundant of these in cell membranes are the phosphatidylcholines, phosphatidylethanolamines and phosphatidylserines, while the phosphatidylinositols and cardiolipins are present in smaller quantities [10]. The sphingolipids have a sphingoid base as their backbone, and these lipids impart the greater hydrophobicity to the core of the lipid bilayer. Cholesterol, on the other hand, has a hydrophilic hydroxyl group, which can interact with the hydrophilic headgroups of the phospholipids, while its bulky steroid group can interact with the hydrophobic acyl chains of the lipids. All of these interactions influence the packing, and thus the fluidity, of the lipid membranes [10,44] (Fig. 2).

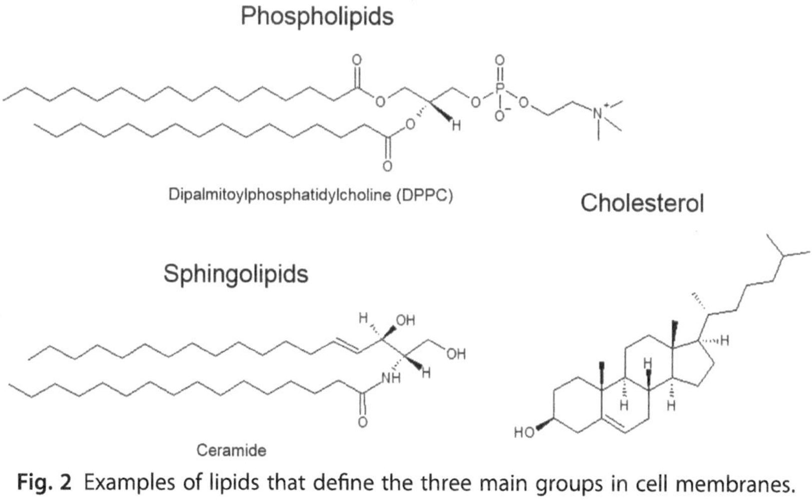

Fig. 2 Examples of lipids that define the three main groups in cell membranes.

As indicated above, on the basis that studies of the properties of specific molecules and their effects on living cells would be too complex, either 'cell ghosts' (i.e., empty cell membranes from living cells) or artificial lipid membranes are usually used instead. However, the creation of a good copy of a biological membrane is relatively complicated; therefore, for *in-vitro* membrane studies, artificial membranes are usually created from specific different types of lipids, such as a simple preparation of liposomes. However, another way of obtaining even better models of biological cell membranes is the creation of cell ghosts by reversible hemolysis of red blood cells, or erythrocytes (i.e., preparation of so-called 'red blood cell ghosts') [10,45,46]. However, although these membrane-enclosed entities are particularly suitable for studying membrane—molecule interactions, artificially created membranes have a much more defined structure and contents. Thus, for the purpose of simplifying the system, lipid-only preparations are generally used. Moreover, as membranes are composed of different kinds of lipids, which can have many different characteristics, artificial membranes are usually made from only one lipid (e.g., dipalmitoylphosphatidylcholine [DPPC]) or a few different lipids (e.g., 1-palmitoyl-2-oleoylphosphatidylcholine [POPC], 1-palmitoyl-2-oleoylphosphatidylethanolamine [POPE], 1-palmitoyl-2-oleoylphosphatidylserine [POPS] and cholesterol). This further simplifies such studies, to allow more decisive conclusions to be made [10,47,48].

There are, however, some drawbacks in the use of pure lipid membranes. It is known that pure lipid bilayers have very high electric resistance (10^6—10^9 Ω/cm^2), and thus there is very low electric conductivity across them. In contrast, the electric conductivity of cell membranes is very high, and their electric resistance is lower by several orders of magnitude (10^3 Ω/cm^2) compared to pure lipid membranes, due to the protein ion channels in cell membranes [49]. Also, as the different phospholipids in cell membranes have different charges, the overall charge associated with artificial lipid membranes would reflect the charge of the lipids from which they are formed [44]. This means, that some highly charged molecules might have trouble interacting with certain types of such membranes, while they would easily interact with others. A similar problem is also encountered due to the lack of membrane proteins in artificial lipid membranes, which means that there is absolutely no active transport across artificial lipid membranes of any kind [4,50]. Effectively, certain molecules can pass easily through lipid membranes (e.g., small neutral molecules, water, larger hydrophobic molecules), while others cannot (e.g., large hydrophilic molecules, charged ions, amino acids, sugars), or would do so only under certain circumstances, such as a change in pH [1,3,4,16,51]. The question that also arises in this case is will the studied compounds interact with the living cell membrane in the same way as, or at least similar to, their interactions with pure lipid bilayers? In this case, it is safe to acknowledge that at least some changes will be noted [52].

Therefore, while studying membrane lipid–compound interactions, some environmental factors need to be strictly defined (e.g., solution pH, solvent used, temperature). First, because the properties of these studied compounds can change (e.g., net charge, solubility), and more importantly, because the properties of the lipid bilayers can also change drastically [51]. Two of the most important parameters mentioned here are certainly the temperature and the solution/solvent. It is generally accepted that fluidity and phase changes of lipids can have considerable effects on membrane transport functions [49,51], and both temperature and solution/solvent can drastically change the properties and state of lipid bilayers.

Lipid bilayers are usually in two different phases, as either disordered (i.e., liquid state), where the lipid chains have greater freedom of movement, or ordered (i.e., gel state), where the lipid chains are more densely packed together. The main gel-to-liquid phase transition of lipids is defined as the T_m, and this generally gives rise to a very intense and sharp peak in the heat capacity (Cp). The T_m corresponds to the transition from the ripple gel intermediate phase (P_β') to the liquid crystalline phase (L_α). On the other hand, the pre-transition temperature, T_{pre}, is usually detected as a much less intense and broader peak at a lower temperature, and this corresponds to the transition from the quasi-lamellar phase (L_β') to the ripple gel intermediate phase (P_β') [53]. The permeability of pure lipid membranes for various compounds also depends on the temperature of the phase transition of the lipids—the permeability is highest at the gel-liquid transition and lowest in the gel state. Otherwise, the T_m of the lipids and its shifts are used to determine certain lipid–molecule interactions [16,51,54,55].

The other very important aspect is the pH of the solution, and consequently the lipid bilayer net charge. As around 20% of cell membrane phospholipids are negatively charged at physiological pH, while positively charged phospholipids are very rare [56], most studies have been performed on lipids that are negatively charged at pH 7 (where physiological pH is 7.4). For this, they have to be made at least partially from the negatively charged phospholipids (like phosphatidylserine, phosphatidylinositol, cardiolipin). This change in pH can thus have substantial effects on the bilayer properties, especially when many charged lipids are used in its formation. If neutral phospholipids are used (like DPPC), small pH changes of the environment have a lesser role in the properties of the lipid bilayer, but they can be crucial for the properties of the compounds used [44,56].

Apart from these drawbacks, working with defined artificial lipid membranes is relatively easy and widespread, and these studies have contributed to our better understanding of the interactions between cell membranes and specific compounds.

3.1 Membrane–molecule interactions

Individual compounds have unique interactions with model lipid membranes, which depend mostly on five factors: polarity of the compound (which is closely connected with its moieties) and its configuration/planarity, the pH of the environment (which is closely connected to its electric charge), the temperature (which is connected to the state of the lipids) and the structure of the artificial membranes (e.g., types of lipids used, presence of cholesterol) [57–59].

Simple molecules are usually polar or non-polar, while larger and more complex ones tend to be both. For example, flavonoids are amphipathic molecules, which means they contain hydrophilic as well as hydrophobic groups and can thus interact with both the hydrophobic cores and the hydrophilic headgroups of the lipid membranes [60]. For flavonoids, the number of OH groups has a major role in the hydrophobicity/hydrophilicity of the molecules, as well as in the number of hydrogen bonds that the compound can form. More OH groups on a molecule define a greater hydrophilic character of the molecule, and thus more interactions with the polar head groups of the lipids [52,54,57,60,61]. For all of the compounds chosen here, their approximate polarities are given as the topological polar surface area, where a higher number means a more polar molecule (Table 1). It was shown that considering cancer Caco-2 cells, molecules with topological surface areas >140 Å^2 permeate slowly or not at all through biological membranes, while those at 60 Å^2 or less permeate quickly through biological membranes [62]. As seen from Table 1, all of the compounds chosen here have topological surface areas <140 Å^2, with the exception of rutin, and can thus at least partially penetrate the membranes. The more hydrophobic a molecule is, the faster its permeation through membranes is [63]. It is also important to note that molecules that can permeate through cell membranes more easily are more bioavailable, and should have greater cytotoxic effects, for example on cancer cells [48,63]. The polarity of

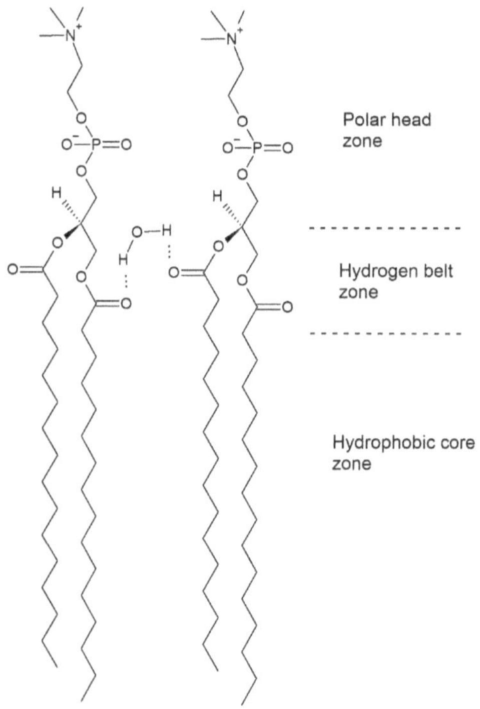

Fig. 3 Structural organization of the outer layer of DPPC phospholipid membranes.

a molecule also depends on its dipole moment, its dielectric constant, and its miscibility with water (number of H-bonds—which is closely connected with the number of OH groups) [64] (Fig. 3).

The electric charge of a molecule is determined by the environment pH, which determines whether the molecule is in a protonated or de-protonated state. In the case of OH groups, these are mostly in the deprotonated state at physiological pH (7.4), and in the protonated state at lower pH (e.g., 3.5). In the case of protonated OH groups (i.e., environment pH, acidic) flavonoids are less polar and are thus incorporated deeper into the hydrophobic core of the membranes, while in their deprotonated state (i.e., environment pH, alkaline) they are negatively charged and more polar, and thus they are incorporated nearer to the polar headgroups of the phospholipids [57,59,65]. It has also been shown that in solutions with higher salt concentrations, flavonoids form stronger interactions with membranes, while when negatively charged, their affinity for membranes is reduced [66].

The conformation of a molecule also has a similar role to that of its polarity. In flavonoids, a double bond between carbon atoms 2 and 3 (C2, C3) in the C ring is responsible for its planar structure. Planar molecules have higher affinities for lipid membranes and perturb the membranes less than non-planar molecules (e.g., naringenin, quercetin), while non-planar compounds penetrate deeper into membranes [57,60,67,68]. As planar molecules have higher affinities for membranes than non-planar ones, their transcellular transport and permeability tend to be lower. For non-planar molecules, it is more likely that these will penetrate through membranes, as it is energetically less favored for them to remain within the membranes than for planar molecules. Furthermore, the conformation of a molecule is more important for its membrane intercalation than its relative hydrophobicity [57,60,69].

Temperature on the other hand has a direct impact on lipid bilayers. If the temperature of the environment is higher than the transition temperature of the phospholipids used, a bilayer will be in its liquid state, while at lower temperatures, the membranes will be in the gel state. If the temperature is the same as the transitional temperature, the bilayers are in the gel-to-liquid-state. This not only influences the interactions of flavonoids with the bilayers, but also the effects that flavonoids will have on the membranes (i.e., more rigid/more fluid membranes) [16,54].

The last factor relates to the types of phospholipids used for membrane creation. As for the temperature, the structure of the phospholipids influences the rigidity of the bilayer. In the case of unsaturated fatty acids in a phospholipid structure (e.g., dioleoylphosphatidylcholine [DOPC]), these make the membrane more fluid, while with saturated fatty acids (e.g., DPPC), the membranes are more rigid. Cholesterol, on the other hand, has similar effects on membranes as flavonoids, and it also tends to intercalate itself into similar places, which often makes it a more difficult for flavonoids to intercalate successfully. Furthermore, different phospholipids have different surface charges. While most are neutral (e.g., phosphatidylcholines), some can have negative charges (e.g., phosphatidylserine), or even positive charges, although rarely. Most flavonoids have a negative charge at physiological pH, so they will be less likely to interact with negatively charged than with neutral membranes, because of their mutual repulsion [16,44,70].

3.2 Phenols–phospholipid interactions

As mentioned above, phenols (and especially flavonoids) are the most important class of compounds in propolis. Therefore, interactions between phenols and model lipid membranes are given the most attention in this

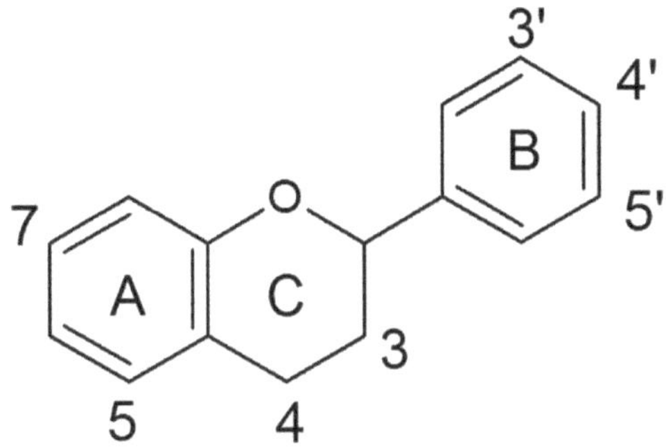

Fig. 4 The basic flavonoid structure.

article, in terms of these propolis constituents. Also, as the phenols have been widely used and studied, there is a lot of available information regarding their interactions with membranes. The phenols discussed here are mostly flavonoids, as can be seen from Table 1. All of these molecules are important propolis constituents with known biological functions, and they have been found in most propolis samples, and especially in poplar propolis (Fig. 4).

One of the most widely studied flavonoids is quercetin. Although its concentrations in propolis are relatively low, it has been defined for most propolis samples. Quercetin has a planar conformation of the double bond between C2 and C3 in the C ring, and it includes five OH groups, which make it one of the most polar constituents among all of these propolis flavonoids. It also includes a ketone group at C4 of the C ring, as for all of these chosen flavonoids [36,37,54]. How quercetin interacts with model lipid membranes remains somewhat controversial, although all studies have reported that it has a planar conformation and interacts strongly with all types of membranes [52,58,68,71,72]. While some studies have argued that quercetin interacts mainly with the polar headgroups of lipids at physiological pH [73–76], others studies have indicated that it predominantly affects the hydrophobic core [77]. It has more recently become generally accepted that quercetin is incorporated in membranes near the border between the hydrophobic acyl chains and the polar lipid headgroups [52,67–69,78–81]. Nonetheless, whether quercetin is incorporated into the membrane core or near to the lipid headgroups is also dependent on the environment pH. At an acidic pH (i.e., in a protonated state), quercetin enters deeper into the hydrophobic part of the membrane, while it interacts mainly with the polar headgroups at physiological pH (i.e., deprotonated state) [65]. There has also been a lot of controversy in terms of whether quercetin makes membranes more fluid or more rigid, although it is generally accepted that flavonoids increase membrane rigidity [82]. Even though some studies have reported rigidifying effects of quercetin on membranes [52,67,77,80,83,84], others have reported that it actually fluidifies them [68,69,79]. It turns out in the end that the rigidifying effects of quercetin are actually determined by two main factors: its concentration in the solution, and hence also in the membranes, and the properties of the membranes themselves. It has been show that higher concentrations of quercetin promote increased membrane rigidity, and lower concentrations promote increased membrane fluidity [76,84,85]. Also, the various studies have used very different membrane models, including (from more simple to complex): monolayer models; liposomes; planar bilayer membranes; saturated and unsaturated membranes; cell ghosts; and living cancer or bacterial cells. It turns out that when membranes are in the liquid state or when they have a high proportion of unsaturated lipids (e.g., POPC, dioleoylphosphatidylethanolamine [DOPE], 1-stearoyl-2-oleoylphosphatidylserine [SOPS], DOPC), quercetin makes them more rigid, while if the membranes are in the gel state or if they are built predominantly from saturated lipids (e.g., DPPC), quercetin makes them more fluid [52,73,74,79,81,83]. This appears to be true for most flavonoids, and it is very similar to the effects of cholesterol on lipid membranes [61,64,86,87]. At the same time, it has also been shown that addition of quercetin has a large impact on the polarization of membranes [58], through promotion of reduced thickness of membranes [69,78], and when at acidic pH, also through increased transmembrane conductance [65]. It has been noted that the effects of quercetin are also greater in the absence of cholesterol [81], and that quercetin undergoes slow transcellular transport, as its five OH groups interact strongly with the membranes [60]. Also, because of its strong interactions and its near-surface position, quercetin is a strong antioxidant and has strong anti-proliferative activity, which is greater than that of other flavonoids [52,76]. Furthermore, when there are very high quercetin concentrations in the membranes, the quercetin penetration into the membranes creates small passages for ions, which leads to the greater ion permeability [65].

The glycoside rutin can be seen to be similar to quercetin. Despite having the same flavonoid body, its effects on membranes are very different because of its sugar moiety, which makes rutin by far the most polar of the selected molecules [36,37]. Due to its size, the interactions of rutin with saturated membranes are much weaker, which reflects its particularly low permeability [68]. Despite its poor binding, rutin is positioned near to the hydrophilic polar headgroup—hydrophobic core border in unsaturated membranes, with its sugar moiety lying outside the membranes [67,69]. In saturated gel-state membranes, rutin binds mainly with the polar headgroups of the lipids [75]. Rutin was shown to rigidify unsaturated membranes [67,88], and similar to quercetin, it also makes them thinner [69].

The other flavonoids chosen here have not been as well studied as quercetin; nonetheless, their activities and interactions with membranes are relatively well documented. Kaempferol and galangin are two flavonoids that are structurally very similar to quercetin, and that also represent an important part of propolis, particularly as galangin is one of the most abundant propolis constituents [33]. These are both planar flavonols and they differ from quercetin only in their number of OH groups on the B ring. Quercetin has two (C3, C4), while kaempferol has one (C4), and galangin has none [36,37]. On this basis, galangin is the most hydrophobic of these three flavonoids. Again, similar to quercetin, kaempferol and galangin have strong impacts on membrane polarization, and they also form strong interactions with membranes. All three of these compounds are intercalated on the border between the hydrophobic core and the polar headgroups [69,81]. As galangin is the most hydrophobic, it is positioned a little deeper into the membranes than quercetin and kaempferol, and it has stronger effects on the membranes [52,58,69,71,76,81]. Galangin has a large impact on proliferation of cancer cells (similar to quercetin) as it forms strong interactions with membranes [52]. As permeability, and in the case of cancer cells, toxicity, are closely related to the polarity of membranes, galangin is the most permeant, followed by kaempferol, and lastly quercetin, which permeates only partially into membranes [63]. Similar to quercetin, kaempferol and galangin also rigidify membranes made of unsaturated lipids [52,64,77], while with DPPC membranes, galangin makes them more fluid [76]. Kaempferol has also been shown to rigidify saturated (i.e., DPPC) membranes when they are in the liquid state [81]. Despite this, it was reported that contrary to quercetin, kaempferol rigidifies unsaturated membranes at lower concentrations, but fluidifies them at higher concentrations. This would appear to be connected to the formation of lipid "rafts" due to the high flavonoid concentrations, which represent regions of membranes with a distinct, characteristic structural composition that appear to act as platforms for colocalization of proteins involved in intracellular signaling pathways [64]. Tsuchiya et al. (2010) also noted that considering quercetin, kaempferol and galangin, of these three flavonols, galangin rigidifies membranes the most and kaempferol the least [52]. Kaempferol was also reported to rigidify artificial bacterial membranes of *Escherichia coli* made from dipalmitoylphosphatidylethanolamine (DPPE) and dipalmitoylphosphatidylglycerol (DPPG) lipids [89]. Also, similar to quercetin, the effects of kaempferol on membranes are greater in the absence of cholesterol [81].

As well as these three flavonols and the flavonol glycoside rutin, some of the most important and abundant propolis constituents also include pinocembrin, a flavanone, and chrysin, luteolin and apigenin, which are flavones. We also investigated the membrane interactions of the flavanone naringenin, because of its similarity to pinocembrin, its strong biological activity, and its presence in propolis. Among this five further propolis constituents, chrysin and pinocembrin are particularly abundant [33]. Chrysin, luteolin and apigenin all have a planar configuration because of the double bond between C2 and C3 of the C ring, and they only differ in terms of the number of OH groups on the B ring: chrysin has none, apigenin has one (C4), and luteolin has two (C3, C4). Unlike the flavonols, chrysin, luteolin and apigenin do not have the OH group on the C4 of the C ring. On the other hand, pinocembrin and naringenin do not have planar configurations like the flavones, as they are both tilted. These two flavanones differ only in their OH groups on the B ring: pinocembrin has none, while naringenin has one on C4. The polarities of all five of these compounds are very similar, although luteolin is a little more polar [36,37]. Compared to the chosen flavonols, the polarization effects of these three flavones, chrysin, luteolin and apigenin, on membranes are weaker, as are their interactions with membranes. The same has been shown for pinocembrin and naringenin. Chrysin, luteolin and apigenin, along with naringenin, are also positioned closer to the polar headgroups in unsaturated membranes, compared to the flavonols [52,57,58,69,90]. Despite this, their membrane affinities are still high, and they are positioned between the polar headgroups and the hydrophobic core [54,67,68,71,84,91,92]. For chrysin, it has been shown to bind well with membranes because of its relative high hydrophobicity, whereby it inserts more deeply into membranes (i.e., with only two OH groups and its planar structure). Contrarywise, luteolin binds well with membranes because of its four OH groups, and it is positioned a little closer to the polar headgroups [60,92,93]. Also, due to their different numbers of OH groups, these two molecules have different orientations in membranes. Chrysin is oriented perpendicular to the plane of the membranes, while luteolin has a parallel orientation [54]. Luteolin, apigenin, chrysin and naringenin have all been shown to rigidify unsaturated membranes, although less so than quercetin and galangin [52,67,80,84,88]. Apigenin was also shown to rigidify saturated membranes in the liquid state [90]. On the other hand, naringenin renders saturated membranes made from DPPC more fluid, as it also does for erythrocyte membranes [68,91]. Similar to quercetin, luteolin does not permeate through membranes well because of its strong affinity to the membranes (i.e., four OH groups and its planar structure) [60].

Phenolic acids interact with lipid membranes in a similar way to flavonoids, although they tend to be more hydrophobic due to their different structures and different numbers of OH groups. The most abundant phenolic acids in propolis are ρ-coumaric, cinnamic, ferulic and caffeic acids. These all include a phenyl group with a prop-2-enoic acid attached to it, which defines the structure of cinnamic acid. ρ-Coumaric acid also includes an OH group at C4 of the phenyl ring, caffeic acid also includes two OH groups at C3 and C4, and ferulic acid also includes an OH group at C4 and a methoxy group at C3. Therefore, cinnamic acid is the most hydrophobic of these phenolic acids, while caffeic acid is the most polar [36,37,48]. Despite having structures that are relatively similar to those of the flavonoids, these phenolic acids have smaller effects on membranes, and their interactions with membranes are weaker. For example, ρ-coumaric and cinnamic acids have very little influence on polarization of membranes in comparison to all three of the flavonols indicated above [58]. Also, caffeic acid has lower affinity for membranes than quercetin [72]. Within these phenolic acids, caffeic and ferulic acids form weaker membrane interactions, while ρ-coumaric acid forms stronger membrane interactions [48]. Nevertheless, they still behave similar to the flavonoids. Caffeic acid, for example, tends to be inserted deeper into the hydrophobic core of membranes in its protonated state (i.e., at acid pH), while it is positioned near to the polar headgroups in its deprotonated state (i.e., at physiological pH), similar to quercetin [78]. Also ferulic, ρ-coumaric and caffeic acids increase the rigidity of unsaturated membranes, while for saturated membranes (i.e., DPPC), ρ-coumaric acid makes them more fluid, with minimal effects reported for caffeic and ferulic acids. For the gel state of membranes, the effects of these phenolic acids are smaller than for the liquid state [48]. It has also been reported that the negatively charged acids cannot permeate through saturated membranes. It is important to note here that almost all of these studies were carried out at physiological pH, where these phenolic acids were in their deprotonated states [48,58,72].

A further molecule worth noting as an important propolis constituent that is closely related to caffeic acid is its phenethyl ester. Although caffeic acid phenethyl ester has been less studied, it has some interesting biological functions. It is an ester of caffeic acid and phenethyl alcohol, which makes it a little more hydrophobic than caffeic acid, with a strong affinity for lipid membranes [36,37,72].

The last three propolis phenols chosen here have particularly different structures compared to the rest of them. The first of these compounds is artepillin C, which is similar in some ways to ρ-coumaric acid, but with two 3-methylbut-2-enyl moieties at C3 and C5 of the phenyl ring. This means that the polarity of artepillin C is similar to that of ferulic acid. The second compound here is α-tocopherol, or vitamin E, which although it is a phenol, it is more similar to the cholesterol molecule. As α-tocopherol includes a long acyl chain and many methyl moieties on its phenyl ring, it is by far the most hydrophobic molecule of these selected phenols. The only molecule with a similar polarity is the volatile phenol eugenol. In contrast to the other phenols, eugenol is found primarily in the "essential oils" fraction of propolis [36,37,94]. Considering artepillin C again, due to its structure it is very amphiphilic and it intercalates well into membranes, while its negative charge lowers the net charge of membranes even more. It also makes membranes thinner, with its intercalation near the border of the polar headgroups and hydrophobic core [95]. On the other hand, α-tocopherol affects membranes in a very similar way to cholesterol. It intercalates deeper into the hydrophobic core due to its acyl tail, and it rigidifies unsaturated membranes [88,96]. However, because of its hydrophilic part, it is also positioned at the border of the polar headgroups and the hydrophobic core [97,98]. α-Tocopherol predominantly intercalates into unsaturated membranes, with which it forms stronger interactions [96], although its diffusion into unsaturated membranes is only a little faster than its diffusion into saturated membranes [99]. Eugenol reacts with membranes in a manner similar to α-tocopherol, and it fluidifies saturated membranes. Eugenol intercalates into membranes near the border between the lipid headgroups and the hydrophobic core in erythrocytes and in DPPC and dipalmitoylphosphatidylserine [DPPS] membranes, while it is positioned in the hydrophobic core with positively charged dioctadecyldimethylammonium bromide [DODAB] membranes. Eugenol was shown to at least partially expand membranes, although it does not have any effects on the permeability of other molecules through erythrocyte membranes [100,101] (Fig. 5).

3.3 Terpenes-phospholipid interactions

The terpenes are another major and biologically very important propolis component. They are volatile and mostly very non-polar compounds. Among the chosen propolis constituents discussed here, the terpenes are far

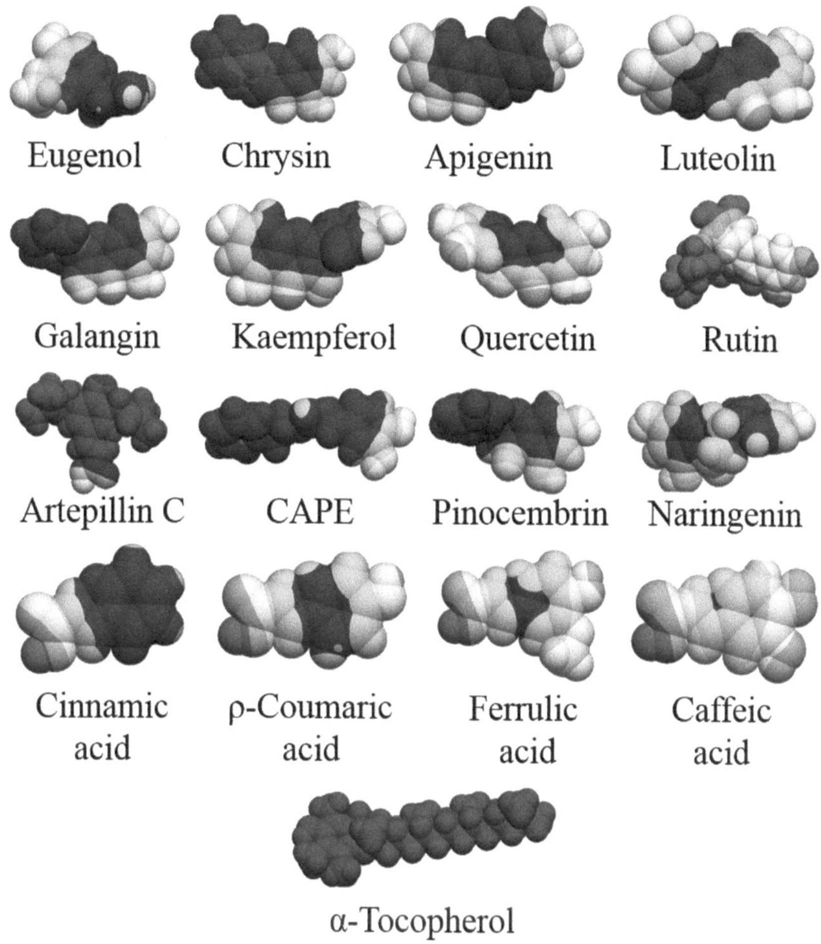

Fig. 5 Lipophilic potential of the propolis phenols discussed here. The darker the color the more is the region of the molecule hydrophobic; CAPE, caffeic acid phenethyl ester. *Figure created with Galaxy Visualizer (Molinspiration Galaxy 3D Structure Generator v2018.01 beta).*

more hydrophobic than the phenols and flavonoids, as seen from Table 1. In propolis, they are mostly found in the 'essential oils' fraction. As terpenes do not have good water solubility, and are thus rarely present in the cellular environment, their interactions with membranes have not been so well studied [29,34,102].

The most non-polar terpene among those chosen here is limonene, a cyclic monoterpene. It is the only terpene that does not include at least one

OH group [36,37,94]. It is completely hydrophobic, and forms small micelles of 18–20 molecules in water solutions. It almost does not react with the polar lipid headgroups, and is instead rapidly incorporated in the hydrophobic core of membranes, where it forms van der Waals interactions with the acyl chains of the lipids. In unsaturated membranes, it is generally positioned near the center of the hydrophobic core [103,104]. In saturated membranes and in fibroblast and erythrocyte membranes, limonene makes the bilayer more fluid [103,105], while in unsaturated membranes, it makes them more rigid, and thicker and denser; the more limonene added, the stronger these effects are [104]. Due to its hydrophobicity, limonene permeates quickly through membranes [104]. It has also been shown that limonene has toxic effects on fibroblast and erythrocyte membranes. Nevertheless, sesquiterpenes are more cytotoxic for membranes than monoterpenes, although they both made membranes more fluid [103].

In contrast to limonene, the terpenes lupeol, α-amyrin, β-amyrin and geraniol all include one OH group, while isocupressic acid includes two. These terpenes chosen here (including limonene) have particularly different structures, although there are some similarities between α-amyrin, β-amyrin and lupeol, which all include at least four C6 rings, and which are all members of the triterpenes class. Among these selected, the monoterpene geraniol is the only one that does not include a carbon ring, while the diterpene acid, isocupressic acid, includes two C6 rings and a long acyl chain [36,37,94]. Lupeol intercalates well into saturated phospholipids in membranes, and also in sphingomyelin membranes, and makes these more fluid. It interacts mostly with the hydrophobic core, as for cholesterol, where it intercalates mostly into the outer layer of lipid membranes. It also has cytotoxic effects on cardiolipin/phospholipid *E. coli* membranes (i.e., degradation, functionality disorders), similar to α-amyrin and β-amyrin. For α-amyrin, it has also been shown to make unsaturated membranes more rigid [106–108]. As the only non-cyclic terpene, geraniol fluidifies saturated membranes and *Candida albicans* membranes, and lowers their surface potential, because of its strong interactions with the lipid acyl chains. Indeed, geraniol expands saturated membranes and make them more permeable for other molecules [109,110]. While most of the terpenes are very hydrophobic, isocupressic acid, on the other hand, is the most polar among the terpenes chosen here. Despite this, isocupressic acid permeates through membranes well, and it is mostly intercalated into the inner layer of erythrocyte bilayers. As it has two OH groups on opposite parts of the molecule, despite its long acyl chain, isocupressic acid does not have the strong effects

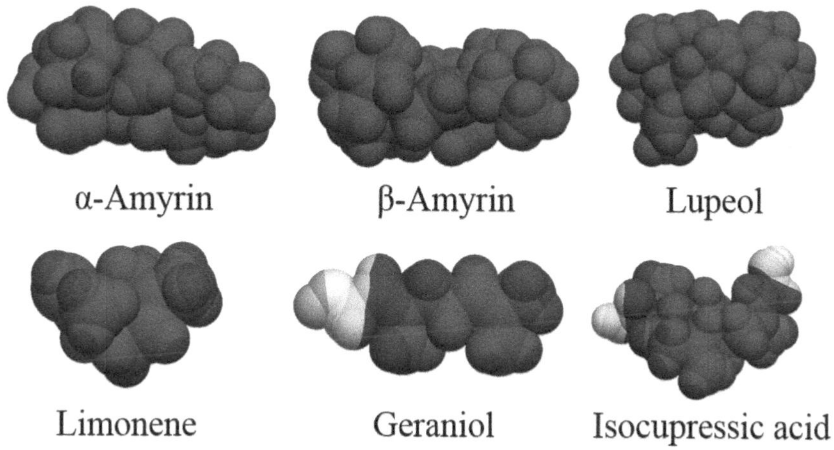

Fig. 6 Lipophilic potential of the chosen terpenes from propolis. The darker the color the more is the region of the molecule hydrophobic. *Figure created with Galaxy Visualizer (Molinspiration Galaxy 3D Structure Generator v2018.01 beta).*

on membranes as seen for these other terpenes. The main reason for this is that isocupressic acid is intercalated closer to the polar headgroups than the other terpenes. Its effects on membranes are also concentration dependent, where higher concentrations have greater effects [111] (Fig. 6).

4. Conclusions

Flavonoids and terpenes are the most biologically important constituents of propolis, and because of their broad spectrum of activities, they have been well studied. Most of these form strong interactions with different types of membranes, from artificial membranes composed of lipids with unsaturated and saturated chains, to cell ghosts, organelle membranes, and bacterial and fungal cells. In general, the flavonoids and terpenes fluidify membranes with saturated chains and membranes with lipids in the gel-state, while they rigidify membranes composed of lipids with unsaturated chains and ones in the liquid state. Their activities are usually concentration dependent. Their interactions with membranes depend largely on their polarity, the numbers and positions of their OH groups (or other moieties), the planarity of the molecules, and the environment pH (for their charge) and temperature (for T_m of the lipids). The uncharged and more hydrophobic flavonoids and terpenes interact more strongly with membranes than the negatively charged and more hydrophilic ones. Hydrophobic and tilted terpenes and flavonoids have also been shown to be incorporated deeper into the hydrophobic core of lipids in membranes. The hydrophilic and planar flavonoids, on the other hand, can form especially strong bonds with the lipid headgroups. Most flavonoids are positioned near the border between the polar headgroups and the hydrophobic core, while most terpenes have higher affinities for the lipid core. The permeability of membranes to these compounds is closely connected to their polarity, whereby the more hydrophobic the molecule, the better its permeability. Many studies have shown that the activities of the flavonoids and terpenes are also linked directly to the type of interactions they have with membranes. Stronger interactions mean stronger biological activities. The binding of the flavonoids and terpenes to membranes usually also expands the membranes and increases the permeability of the membranes to other molecules.

Nevertheless, despite many individual studies on flavonoids and terpenes and their interactions with membranes, studies that have focused on their synergistic effects on membranes are scarce. Therefore, future studies should also consider the synergistic effects that these compounds can have on model membranes, particularly as it has already been shown that for propolis, synergistic effects are stronger than the individual effects of these compounds.

References

[1] G.M. Cooper, R.E. Hausman, The Cell: A Molecular Approach, fourth ed., ASM Press, Washington, DC, Sunderland, MA, 2007.

[2] R.B. Gennis, Biomembranes: Molecular Structure and Function, first ed., Springer-Verlag, New York, 1989.

[3] H.F. Lodish, A. Berk, C. Kaiser, M. Krieger, M.P. Scott, A. Bretscher, H.L. Ploegh, P.T. Matsudaira, Molecular Cell Biology, W.H. Freeman, New York, 2008.

[4] D.E. Sadava, E.C. Heller, W.K. Purves, G.H. Orians, D.M. Hillis, Life: The Science of Biology, eighth ed., Sinauer Associates Inc. and W. H. Freeman & Co., Sunderland, MA and Gordonsville, VA, U.S.A, 2008.

[5] H.E. Grecco, M. Schmick, P.I.H. Bastiaens, Signaling from the living plasma membrane, Cell 144 (6) (2011) 897–909.

[6] H. Sunshine, M.L. Iruela-Arispe, Membrane lipids and cell signaling, Curr. Opin. Lipidol. 28 (5) (2017) 408–413.

[7] I. Ghai, S. Ghai, Understanding antibiotic resistance via outer membrane permeability, Infect. Drug Resist. 11 (2018) 523–530.

[8] G. Balogh, M. Péter, A. Glatz, I. Gombos, Z. Török, I. Horváth, J.L. Harwood, L. Vígh, Key role of lipids in heat stress management, FEBS Lett. 587 (13) (2013) 1970–1980.

[9] S. Kaye, S. Merry, Tumour cell resistance to anthracyclines - a review, Canc. Chemother. Pharmacol. 14 (2) (1985) 96–103.

[10] C. Peetla, A. Stine, V. Labhasetwar, Biophysical interactions with model lipid membranes: applications in drug discovery and drug delivery, Mol. Pharm. 6 (5) (2009) 1264–1276.

[11] K.D. Taylor, R. Goel, F.H. Shirazi, M. Molepo, P. Popovic, D.J. Stewart, P.T. Wong, Pressure tuning infrared spectroscopic study of cisplatin-induced structural changes in a phosphatidylserine model membrane, Brit. J. Cancer 72 (6) (1995) 1400–1405.
[12] E. Alday, M. Navarro-Navarro, A. Garibay-Escobar, R. Robles-Zepeda, J. Hernandez, C. Velazquez, in: E.D. Chambo (Ed.), Beekeeping and Bee Conservation - Advances in Research, IntechOpen, 2016, https://doi.org/10.5772/63145.
[13] M.N. Kardar, T. Zhang, G.D. Coxon, D.G. Watson, J. Fearnley, V. Seidel, Characterisation of triterpenes and new phenolic lipids in Cameroonian propolis, Phytochemistry 106 (2014) 156–163.
[14] J.M. Sforcin, Biological properties and therapeutic applications of propolis, Phytother. Res. 30 (6) (2016) 894–905.
[15] S. Castro, Propolis: biological and pharmacological activities. Therapeutic uses of this bee product, Annu. Rev. Biomed. Sci. 3 (2001) 49–83.
[16] C.G. Siontorou, G.P. Nikoleli, D.P. Nikolelis, S.K. Karapetis, Artificial lipid membranes: past, present, and future, Membranes 7 (3) (2017) 38.
[17] A.B. Hendrich, Flavonoid-membrane interactions: possible consequences for biological effects of some polyphenolic compounds, Acta Pharmacol. Sin. 27 (1) (2006) 27–40.
[18] G.A. Burdock, Review of the biological properties and toxicity of bee propolis (propolis), Food Chem. Toxicol. 36 (4) (1998) 347–363.
[19] E.L. Ghisalberti, Propolis: a review, Bee World 60 (2) (1979) 59–84.
[20] A.K. Kuropatnicki, E. Szliszka, W. Krol, Historical aspects of propolis research in modern times, Evid. Based Compl. Alt. 2013 (2013) 11.
[21] M.C. Marcucci, Propolis: chemical composition, biological properties and therapeutic activity, Apidologie 26 (2) (1995) 83–99.
[22] M.G. Miguel, A.C. Figueiredo, in: J.M. Alvarez-Suarez (Ed.), Bee Products - Chemical and Biological Properties, Springer International Publishing AG, Cham, Switzerland, 2017, p. 306, ch. 6.
[23] V.D. Wagh, Propolis: a wonder bees product and its pharmacological potentials, Adv. Pharmaceut. Sci. 2013 (2013) 11.
[24] S. Alves de Souza, C.A. Camara, E. Monica Sarmento da Silva, T.M. Silva, Composition and antioxidant activity of geopropolis collected by *Melipona subnitida* (Jandaira) bees, Evid. Based Compl. Alt. 2013 (2013) 5.
[25] K. Graikou, M. Popova, O. Gortzi, V. Bankova, I. Chinou, Characterization and biological evaluation of selected Mediterranean propolis samples. Is it a new type? LWT - Food Sci. Technol. 65 (2016) 261–267.
[26] J. Zhang, X. Shen, K. Wang, X. Cao, C. Zhang, H. Zheng, F. Hu, Antioxidant activities and molecular mechanisms of the ethanol extracts of *Baccharis* propolis and *Eucalyptus* propolis in RAW64.7 cells, Pharm. Biol. 54 (10) (2016) 2220–2235.
[27] V.S. Bankova, S.L. de Castro, M.C. Marcucci, Propolis: recent advances in chemistry and plant origin, Apidologie 31 (1) (2000) 3–15.
[28] H. Fokt, A. Pereira, A.M. Ferreira, A. Cunha, C. Aguiar, in: A. Méndez-Villas (Ed.), Current Research, Technology and Education Topics in Applied Microbiology and Microbial Biotechnology, vol. 1, Formatex Research Center, 2010, pp. 481–493.
[29] S. Huang, C.P. Zhang, K. Wang, G.Q. Li, F.L. Hu, Recent advances in the chemical composition of propolis, Molecules 19 (12) (2014) 19610–19632.
[30] P. Walker, E. Crane, Constituents of propolis, Apidologie 18 (4) (1987) 327–334.
[31] C.P. Zhang, G. Liu, F.L. Hu, Hydrolysis of flavonoid glycosides by propolis β-glycosidase, Nat. Prod. Res. 26 (3) (2012) 270–273.
[32] V. Bankova, Chemical diversity of propolis and the problem of standardization, J. Ethnopharmacol. 100 (1–2) (2005) 114–117.

[33] C. Gardana, M. Scaglianti, P. Pietta, P. Simonetti, Analysis of the polyphenolic fraction of propolis from different sources by liquid chromatography-tandem mass spectrometry, J. Pharmaceut. Biomed. 45 (3) (2007) 390−399.
[34] V. Bankova, M. Popova, B. Trusheva, Propolis volatile compounds: chemical diversity and biological activity: a review, Chem. Cent. J. 8 (1) (2014) 8.
[35] M.P. Popova, I.B. Chinou, I.N. Marekov, V.S. Bankova, Terpenes with antimicrobial activity from Cretan propolis, Phytochemistry 70 (10) (2009) 1262−1271.
[36] A. Gindulyte, B.A. Shoemaker, B. Yu, J. He, J. Zhang, J. Chen, L. Zaslavsky, P.A. Thiessen, Q. Li, S. He, S. Kim, T. Cheng, E.E. Bolton, PubChem 2019 update: improved access to chemical data, Nucleic Acids Res. 47 (D1) (2018) D1102−D1109.
[37] Pubchemncbinlmnihgov, Topological polar surface area, https://pubchem.ncbi.nlm.nih.gov. (Accessed 25 March 2019).
[38] A.H. Banskota, Y. Tezuka, S. Kadota, Recent progress in pharmacological research of propolis, Phytother. Res. 15 (7) (2001) 561−571.
[39] O. Cuesta-Rubio, M. Campo Fernandez, I. Marquez Hernandez, C.G.J. Jaramillo, V.H. Gonzalez, R. Montes De Oca Porto, D. Marrero Delange, L. Monzote Fidalgo, A.L. Piccinelli, L. Campone, L. Rastrelli, Chemical profile and anti-leishmanial activity of three Ecuadorian propolis samples from Quito, Guayaquil and Cotacachi regions, Fitoterapia 120 (2017) 177−183.
[40] S. Kumazawa, T. Hamasaka, T. Nakayama, Antioxidant activity of propolis of various geographic origins, Food Chem. 84 (3) (2004) 329−339.
[41] A. Mavri, H. Abramovič, T. Polak, J. Bertoncelj, P. Jamnik, S. Smole Možina, B. Jeršek, Chemical properties and antioxidant and antimicrobial activities of Slovenian propolis, Chem. Biodivers. 9 (8) (2012) 1545−1558.
[42] P. Ristivojevic, J. Trifkovic, U. Gasic, F. Andric, N. Nedic, Z. Tesic, D. Milojkovic-Opsenica, Ultrahigh-performance liquid chromatography and mass spectrometry (UHPLC-LTQ/Orbitrap/MS/MS) study of phenolic profile of Serbian poplar type propolis, Phytochem. Anal. 26 (2) (2015) 127−136.
[43] S. Almutairi, B. Eapen, S.M. Chundi, A. Akhalil, W. Siheri, C. Clements, J. Fearnley, D.G. Watson, R. Edrada-Ebel, New anti-trypanosomal active prenylated compounds from African propolis, Phytochem. Lett. 10 (2014) 35−39.
[44] J. Li, X. Wang, T. Zhang, C. Wang, Z. Huang, X. Luo, Y. Deng, A review on phospholipids and their main applications in drug delivery systems, Asian J. Pharm. Sci. 10 (2) (2015) 81−98.
[45] J.F. Hoffman, Physiological characteristics of human red blood cell ghosts, J. Gen. Physiol. 42 (1) (1958) 9−28.
[46] T.J. Simons, The preparation of human red cell ghosts containing calcium buffers, J. Physiol. 256 (1) (1976) 209−225.
[47] M. Edidin, Lipids on the frontier: a century of cell-membrane bilayers, Nat. Rev. Mol. Cell Biol. 4 (5) (2003) 414−418.
[48] A. Ota, H. Abramovič, V. Abram, N. Poklar Ulrih, Interactions of p-coumaric, caffeic and ferulic acids and their styrenes with model lipid membranes, Food Chem. 125 (4) (2011) 1256−1261.
[49] I. Zs-Nagy, The role of membrane structure and function in cellular aging: a review, Mech. Ageing Dev. 9 (3) (1979) 237−246.
[50] B. Alberts, A. Johnson, J. Lewis, M. Raff, K. Roberts, P. Walter, Molecular Biology of the Cell, fourth ed., Garland Science, 2002.
[51] F. Frézard, Liposomes: from biophysics to the design of peptide vaccines, Braz, J. Med. Biol. Res. 32 (2) (1999) 181−189.
[52] H. Tsuchiya, Structure-dependent membrane interaction of flavonoids associated with their bioactivity, Food Chem. 120 (4) (2010) 1089−1096.

[53] K.A. Riske, R.P. Barroso, C.C. Vequi-Suplicy, R. Germano, V.B. Henriques, M.T. Lamy, Lipid bilayer pre-transition as the beginning of the melting process, Biochim. Biophys. Acta 1788 (5) (2009) 954–963.
[54] R. Sinha, A. Joshi, U.J. Joshi, S. Srivastava, G. Govil, Localization and interaction of hydroxyflavones with lipid bilayer model membranes: a study using DSC and multinuclear NMR, Eur. J. Med. Chem. 80 (2014) 285–294.
[55] R.E. Pagano, R.J. Cherry, D. Chapman, Phase transitions and heterogeneity in lipid bilayers, Science 181 (4099) (1973) 557–559.
[56] S. McLaughlin, in: F. Bronner, A. Kleinzeller (Eds.), Current Topics in Membranes and Transport, vol. 9, Academic Press, 1977, pp. 71–144.
[57] C. van Dijk, A.J.M. Driessen, K. Recourt, The uncoupling efficiency and affinity of flavonoids for vesicles, Biochem. Pharmacol. 60 (11) (2000) 1593–1600.
[58] A.G. Erlejman, S.V. Verstraeten, C.G. Fraga, P.I. Oteiza, The interaction of flavonoids with membranes: potential determinant of flavonoid antioxidant effects, Free Radic. Res. 38 (12) (2004) 1311–1320.
[59] F. Ollila, K. Halling, P. Vuorela, H. Vuorela, J.P. Slotte, Characterization of flavonoid–biomembrane interactions, Arch. Biochem. Biophys. 399 (1) (2002) 103–108.
[60] P. Tammela, L. Laitinen, A. Galkin, T. Wennberg, R. Heczko, H. Vuorela, J.P. Slotte, P. Vuorela, Permeability characteristics and membrane affinity of flavonoids and alkyl gallates in Caco-2 cells and in phospholipid vesicles, Arch. Biochem. Biophys. 425 (2) (2004) 193–199.
[61] V. Abram, B. Berlec, A. Ota, M. Sentjurc, P. Blatnik, N.P. Ulrih, Effect of flavonoid structure on the fluidity of model lipid membranes, Food Chem. 139 (1–4) (2013) 804–813.
[62] H. Pajouhesh, G.R. Lenz, Medicinal chemical properties of successful central nervous system drugs, NeuroRx, J. Amer. Soc. Exper. NeuroTherapeut. 2 (4) (2005) 541–553.
[63] K. Kajiya, M. Ichiba, M. Kuwabara, S. Kumazawa, T. Nakayama, Role of lipophilicity and hydrogen peroxide formation in the cytotoxicity of flavonols, Biosc. Biotech. Biochem. 65 (5) (2001) 1227–1229.
[64] N. Poklar Ulrih, A. Ota, M. Šentjurc, S. Kure, V. Abram, Flavonoids and cell membrane fluidity, Food Chem. 121 (1) (2010) 78–84.
[65] L. Movileanu, I. Neagoe, M.L. Flonta, Interaction of the antioxidant flavonoid quercetin with planar lipid bilayers, Int. J. Pharm. (Amst.) 205 (1) (2000) 135–146.
[66] K. Kajiya, S. Kumazawa, T. Nakayama, Effects of external factors on the interaction of tea catechins with lipid bilayers, Biosc. Biotech. Biochem. 66 (11) (2002) 2330–2335.
[67] A. Tedeschi, G. D'Errico, M.R. Lauro, F. Sansone, S. Di Marino, A.M. D'Ursi, R.P. Aquino, Effect of flavonoids on the Aβ (25-35)-phospholipid bilayers interaction, Eur. J. Med. Chem. 45 (9) (2010) 3998–4003.
[68] A. Saija, M. Scalese, M. Lanza, D. Marzullo, F. Bonina, F. Castelli, Flavonoids as antioxidant agents: importance of their interaction with biomembranes, Free Radical Biol. Med. 19 (4) (1995) 481–486.
[69] D. Sanver, B.S. Murray, A. Sadeghpour, M. Rappolt, A.L. Nelson, Experimental modeling of flavonoid–biomembrane interactions, Langmuir 32 (49) (2016) 13234–13243.
[70] N. Poklar Ulrih, A. Ota, V. Abram, Impact of selected polyphenolics on the structural properties of model lipid membranes - a review, Int. J. Food Stud. 6 (2017) 159–177.
[71] K. Murota, A. Obata, M. Kikuchi, T. Izumi, S. Miyamoto, S. Shimizu, J. Terao, Unique uptake and transport of isoflavone aglycones by human intestinal Caco-2 cells: comparison of isoflavonoids and flavonoids, J. Nutr. 132 (7) (2002) 1956–1961.

[72] T. Nakayama, K. Ono, K. Hashimoto, Affinity of antioxidative polyphenols for lipid bilayers evaluated with a liposome system, Biosc. Biotech. Biochem. 62 (5) (1998) 1005–1007.
[73] B. Pawlikowska-Pawlega, W.I. Gruszecki, L. Misiak, R. Paduch, T. Piersiak, B. Zarzyka, J. Pawelec, A. Gawron, Modification of membranes by quercetin, a naturally occurring flavonoid, via its incorporation in the polar head group, Biochim. Biophys. Acta 1768 (9) (2007) 2195–2204.
[74] B. Pawlikowska-Pawlęga, W.I. Gruszecki, L.E. Misiak, A. Gawron, The study of the quercetin action on human erythrocyte membranes, Biochem. Pharmacol. 66 (4) (2003) 605–612.
[75] A.K. Ratty, J. Sunamoto, N.P. Das, Interaction of flavonoids with 1,1-diphenyl-2-picrylhydrazyl free radical, liposomal membranes and soybean lipoxygenase-1, Biochem. Pharmacol. 37 (6) (1988) 989–995.
[76] R. Sinha, S. Srivastava, A. Joshi, U.J. Joshi, G. Govil, *In-vitro* anti-proliferative and anti-oxidant activity of galangin, fisetin and quercetin: role of localization and intermolecular interaction in model membrane, Eur. J. Med. Chem. 79 (2014) 102–109.
[77] D.J. Dorta, A.A. Pigoso, F.E. Mingatto, T. Rodrigues, I.M. Prado, A.F. Helena, S.A. Uyemura, A.C. Santos, C. Curti, The interaction of flavonoids with mitochondria: effects on energetic processes, Chem. Biol. Interact. 152 (2–3) (2005) 67–78.
[78] P. Kosinova, K. Berka, M. Wykes, M. Otyepka, P. Trouillas, Positioning of antioxidant quercetin and its metabolites in lipid bilayer membranes: implication for their lipid-peroxidation inhibition, J. Phys. Chem. B 116 (4) (2012) 1309–1318.
[79] B. Pawlikowska-Pawlega, H. Dziubinska, E. Krol, K. Trebacz, A. Jarosz-Wilkolazka, R. Paduch, A. Gawron, W.I. Gruszecki, Characteristics of quercetin interactions with liposomal and vacuolar membranes, Biochim. Biophys. Acta 1838 (1 Pt B) (2014) 254–265.
[80] H. Tsuchiya, Effects of red wine flavonoid components on biomembranes and cell proliferation, Int. J. Wine Res. 2011 (3) (2011) 9–17, https://doi.org/10.2147/ijwr.s19033.
[81] N.P. Ulrih, M. Maričić, A. Ota, M. Šentjurc, V. Abram, Kaempferol and quercetin interactions with model lipid membranes, Food Res. Int. 71 (2015) 146–154.
[82] A.A.-F. Lenne-Gouverneur, Interactions of the monomeric and dimeric flavones apigenin and amentoflavone with the plasma membrane of L929 cells; a fluorescence study, Mol. Membr. Biol. 16 (2) (1999) 157–165.
[83] D. Margina, M. Ilie, G. Manda, I. Neagoe, M. Mocanu, D. Ionescu, D. Gradinaru, C. Ganea, Quercetin and epigallocatechin gallate effects on the cell membranes biophysical properties correlate with their antioxidant potential, Gen. Physiol. Biophys. 31 (1) (2012) 47–55.
[84] H. Tsuchiya, M. Nagayama, T. Tanaka, M. Furusawa, M. Kashimata, H. Takeuchi, Membrane-rigidifying effects of anti-cancer dietary factors, Biofactors 16 (3-4) (2002) 45–56.
[85] A. Sadeghpour, D. Sanver, M. Rappolt, in: A. Iglič, A. Garcia-Sáez, M. Rappolt (Eds.), Advances in Biomembranes an Lipid Self-Assembly, vol. 25, Elsevier, London, 2017, pp. 95–123.
[86] A. Kusumi, W.K. Subczynski, M. Pasenkiewicz-Gierula, J.S. Hyde, H. Merkle, Spin-label studies on phosphatidylcholine-cholesterol membranes: effects of alkyl chain length and unsaturation in the fluid phase, Biochim. Biophys. Acta Biomembr. 854 (2) (1986) 307–317.
[87] T.P.W. McMullen, R.N.A.H. Lewis, R.N. McElhaney, Differential scanning calorimetric study of the effect of cholesterol on the thermotropic phase behavior of a homologous series of linear saturated phosphatidylcholines, Biochemistry 32 (2) (1993) 516–522.

[88] A. Arora, T.M. Byrem, M.G. Nair, G.M. Strasburg, Modulation of liposomal membrane fluidity by flavonoids and isoflavonoids, Arch. Biochem. Biophys. 373 (1) (2000) 102–109.
[89] M. He, T. Wu, S. Pan, X. Xu, Antimicrobial mechanism of flavonoids against *Escherichia coli* ATCC 25922 by model membrane study, Appl. Surf. Sci. 305 (2014) 515–521.
[90] B. Pawlikowska-Pawlega, L.E. Misiak, B. Zarzyka, R. Paduch, A. Gawron, W.I. Gruszecki, FTIR, (1)H NMR and EPR spectroscopy studies on the interaction of flavone apigenin with dipalmitoylphosphatidylcholine liposomes, Biochim. Biophys. Acta 1828 (2) (2013) 518–527.
[91] V. Ajdzanovic, V. Jakovljevic, D. Milenkovic, A. Konic-Ristic, J. Zivanovic, I. Jaric, V. Milosevic, Positive effects of naringenin on near-surface membrane fluidity in human erythrocytes, Acta Physiol. Hung. 102 (2) (2015) 131–136.
[92] S. Selvaraj, S. Krishnaswamy, V. Devashya, S. Sethuraman, U.M. Krishnan, Investigations on membrane perturbation by chrysin and its copper complex using self-assembled lipid bilayers, Langmuir 27 (21) (2011) 13374–13382.
[93] H.A. Scheidt, A. Pampel, L. Nissler, R. Gebhardt, D. Huster, Investigation of the membrane localization and distribution of flavonoids by high-resolution magic angle spinning NMR spectroscopy, Biochim. Biophys. Acta 1663 (1–2) (2004) 97–107.
[94] S. Clarke, in: S. Clarke (Ed.), Essential Chemistry for Aromatherapy, second ed., Churchill Livingstone, Edinburgh, 2008, pp. 41–77, https://doi.org/10.1016/B978-0-443-10403-9.00003-0.
[95] W.M. Pazin, D.D. Olivier, N. Vilanova, A.P. Ramos, I.K. Voets, A.E. Soares, A.S. Ito, Interaction of artepillin C with model membranes, Eur. Biophys. J. 46 (4) (2017) 383–393.
[96] B. Maggio, A.T. Diplock, J.A. Lucy, Interactions of tocopherols and ubiquinones with monolayers of phospholipids, Biochem. J. 161 (1) (1977) 111–121.
[97] B. Perly, I.C.P. Smith, L. Hughes, G.W. Burton, K.U. Ingold, Estimation of the location of natural α-tocopherol in lipid bilayers by 13C-NMR spectroscopy, Biochim. Biophys. Acta Biomembr. 819 (1) (1985) 131–135.
[98] J. Salgado, J. Villalain, J.C. Gomezfernandez, α-Tocopherol interacts with natural micelle-forming single-chain phospholipids stabilizing the bilayer phase, Arch. Biochem. Biophys. 306 (2) (1993) 368–376.
[99] R. Meyer, A.F. Sonnen, W.M. Nau, Phase-dependent lateral diffusion of α-tocopherol in DPPC liposomes monitored by fluorescence quenching, Langmuir 26 (18) (2010) 14723–14729.
[100] G.E. Goncalves, F.S. de Souza, J.H. Lago, L. Caseli, The interaction of eugenol with cell membrane models at the air-water interface is modulated by the lipid monolayer composition, Biophys. Chem. 207 (2015) 7–12.
[101] G.N. Reiner, L. Delgado-Marin, N. Olguin, S. Sanchez-Redondo, M. Sanchez-Borzone, E. Rodriguez-Farre, C. Sunol, D.A. Garcia, GABAergic pharmacological activity of propofol related compounds as possible enhancers of general anesthetics and interaction with membranes, Cell Biochem. Biophys. 67 (2) (2013) 515–525.
[102] M.A.R. Martins, L.P. Silva, O. Ferreira, B. Schröder, J.A.P. Coutinho, S.P. Pinho, Terpenes solubility in water and their environmental distribution, J. Mol. Liq. 241 (2017) 996–1002.
[103] S.A. Mendanha, S.S. Moura, J.L. Anjos, M.C. Valadares, A. Alonso, Toxicity of terpenes on fibroblast cells compared to their hemolytic potential and increase in erythrocyte membrane fluidity, Toxicol. Vitro 27 (1) (2013) 323–329.
[104] S. Witzke, L. Duelund, J. Kongsted, M. Petersen, O.G. Mouritsen, H. Khandelia, Inclusion of terpenoid plant extracts in lipid bilayers investigated by molecular dynamics simulations, J. Phys. Chem. B 114 (48) (2010) 15825–15831.

[105] S.A. Mendanha, A. Alonso, Effects of terpenes on fluidity and lipid extraction in phospholipid membranes, Biophys. Chem. 198 (2015) 45–54.
[106] M. Broniatowski, M. Flasinski, P. Wydro, Investigation of the interactions of lupane type pentacyclic triterpenes with outer leaflet membrane phospholipids–Langmuir monolayer and synchrotron X-ray scattering study, J. Colloid Interface Sci. 381 (1) (2012) 116–124.
[107] M. Broniatowski, M. Flasinski, K. Zieba, P. Miskowiec, Langmuir monolayer studies of the interaction of monoamphiphilic pentacyclic triterpenes with anionic mitochondrial and bacterial membrane phospholipids - searching for the most active terpene, Biochim. Biophys. Acta 1838 (10) (2014) 2460–2472.
[108] M. Broniatowski, P. Mastalerz, M. Flasinski, Studies of the interactions of ursane-type bioactive terpenes with the model of *Escherichia coli* inner membrane-Langmuir monolayer approach, Biochim. Biophys. Acta 1848 (2) (2015) 469–476.
[109] M. Bard, M.R. Albrecht, N. Gupta, C.J. Guynn, W. Stillwell, Geraniol interferes with membrane functions in strains of *Candida* and *Saccharomyces*, Lipids 23 (6) (1988) 534–538.
[110] S.W. Nowotarska, K.J. Nowotarski, M. Friedman, C. Situ, Effect of structure on the interactions between five natural antimicrobial compounds and phospholipids of bacterial cell membrane on model monolayers, Molecules 19 (6) (2014) 7497–7515.
[111] J. Asili, M. Lambert, H.L. Ziegler, M. Sairafianpour, M. Witt, G. Asghari, I.S. Ibrahimi, J.W. Jaroszewski, Labdanes and isopimaranes from *Platycladus orientalis* and their effects on erythrocyte membrane and on *Plasmodium falciparum* growth in the erythrocyte host cells, J. Nat. Prod. 67 (4) (2004) 631–637.
[112] L. Šturm, N. Poklar Ulrih, Advances in the propolis chemical composition between 2013 and 2018, A review eFood 1 (1) (2020) 24–37, https://doi.org/10.2991/efood.k.191029.001, eISSN 2666-3066 In this issue.

CHAPTER THREE

The human EV membranome

Ana Paulina Ramos Juarez[a,b], Francesco Trepiccione[b], Giovambattista Capasso[b] and Gabriella Pocsfalvi[a,*]

[a]Extracellular Vesicles and Mass Spectrometry Laboratory, Institute of Biosciences and BioResources (IBBR), National Research Council of Italy (CNR), Naples, Italy
[b]First Division of Nephrology, Department of Clinical and Experimental Medicine, School of Medicine, University of Campania Luigi Vanvitelli, Naples, Italy
[*]Corresponding author E-mail: gabriella.pocsfalvi@ibbr.cnr.it

Contents

Abstract

Extracellular vesicles (EVs) are membrane bound structures that gained attention over the last decades because they mediate the horizontal transfer of their cargo molecules.

Advances in Biomembranes and Lipid Self-Assembly, Volume 32
ISSN 2451-9634
https://doi.org/10.1016/bs.abl.2020.09.002

Here, we have built the human EV membranome database by the combination of two *in silico* datasets obtained by: (i) GO term analysis of the human EV proteome and (ii) the extraction of proteins detected in EV preparations from the human membranome database. The EV membranome database contains 1299 frequently identified EV membrane proteins (MPs) and was used for enrichment analysis as well as for data mining of proteins belonging to relevant functional classes. We show that the majority of the EV MPs have plasma membrane origin. However, the presence of MPs of late-endosomal origin and the enrichment of exocytosis related GO term confirms the endosomal route as an alternative way to plasma membrane budding for the secretion of certain EV populations. Enrichment in proteins binding RNA, lipid and carbohydrate highlights the role of EV MPs as signals or ligands in intercellular communication. There are 18 tetraspanins and 40 glycophosphatidylinositol-anchored proteins in the EV membranome that have been frequently identified in EV isolates. Analysis of EV proteome revealed several G-protein-coupled receptor binding proteins (25 proteins) implicated in EV uptake and a high number of transmembrane receptor and tyrosine kinases signaling pathways related surface receptors (80 proteins) involved in cellular communication. These functional classes of EV MPs can represent important pharmaceuticals and clinical targets.

1. Introduction

1.1 Biomembranes and biomembrane components

Biomembranes are selectively permeable structures that form the outer boundary of a cell or a cell compartment. They can be found in all living organisms and are known to have important roles in cellular processes, like active and passive transport, signaling, cell recognition, and compartmentalization. The cell membrane or plasma membrane separates the cell interior from the extracellular milieu and it is typically 4–10 nm in thickness. The primary role of the plasma membrane is to maintain cellular integrity. For example, archaeal extremophiles live in a harsh environment and due to their cell membranes, they can withstand low pH, high temperatures, the presence of toxic metals, *etc.* Besides maintaining cellular integrity and shape, molecular components of plasma membranes are involved in the passive and active molecular transport [1]. The biomembrane defines the outer limit not only of the cell but also of many internal organelles such as nucleus, mitochondria, lysosome, endosome, vacuole, Golgi, intra and extracellular vesicles. Organelle membranes are usually a single layer of lipids. The nucleus, mitochondria, chloroplasts and vesicles are the exception and have a lipid bilayer. The biomembrane of each organelle has a unique and dynamic molecular composition that allows to fulfill different cellular roles from protein synthesis to cellular communication. For example, the mitochondrion has two membranes that differ in composition: a smooth porous outer membrane with proteins that allows the movements of cytosolic proteins and ions into the intermembrane space, and an inner membrane, which is highly folded, much less permeable and contains different highly specific transport proteins that regulate the entrance of substances.

The main molecular components of biomembranes are lipids, proteins and carbohydrates. Lipids represent around 40% of the total mass composition of biomembranes [2]. There are hundreds of different lipid species that are distributed asymmetrically between the internal and external leaflets of the biomembrane. The most abundant lipids in biomembranes are phospholipids, however, there are other amphipathic lipids such as sphingolipids and sterols, which highly differ in their chemical structure and confer different characteristics to the membrane. Lipid composition influences both the thickness and the fluidity of a biological membrane and thus, is associated with the membrane molecular dynamics including lateral organization of proteins [1].

Membrane proteins (MPs) contribute about 50% of the total mass of a typical mammalian plasma membrane with a protein-to-lipid number ratio of about 1:50 [3]. One-third of the human proteome is predicted to be membrane-associated which shows that biomembranes are indeed highly crowded in proteins [4,5]. MPs function as enzymes, transporters, receptors, pores, *etc.* The composition of proteins lends unique properties to the biomembrane [6]. It is important to note, that MPs represent more than 60% of pharmaceutical targets [7] today. The number of identified and structurally characterized MPs are rapidly increasing with the advances of omics, imaging and computational sciences. Different membranome databases, from more general, like Protein Data Bank (PDB) [8] and UniProtKB [9] to more specific ones like Orientations of Proteins in Membranes (OPM) [10], Topology Database of Transmembrane Proteins (TOPDB) [11], Transporter Classification Database (TCDB) [12], Membranome 2.0 database [13], Human Transmembrane Proteome (HTP) [14] and *in silico* human surfaceome [15,16] are available today to assist the analysis and computational modeling. Some of them collect MPs with specific functions, such as receptors, transporters, ion channels, *etc.* [12,13,15] and some others are specific for a given organism [13,17]. In this work, we have used the recently published human surfaceome database (http://wlab.ethz.ch/surfaceome/) developed with a machine learning approach using domain specific features, contains both *in silico* membranome (5102 proteins) and surfaceome (2886 proteins), which is a subset of the human membranome [15]. Surfaceome database provides functional and structural information on membrane proteins, links to UniProt database, number of transmembrane domains, topology, subcellular annotation, drug bank approved IDs and gene IDs.

A broad classification for MPs was proposed by Simons and Nicolson back in the '70s based on the degree of association of the protein to the biomembranes: peripheral proteins, membrane associated proteins and integral proteins. Peripheral MPs are weakly bound to the membrane by electrostatic interactions and when dissociate they are relatively soluble. Peripheral MPs can be attached to the lipid or to the protein part of the biomembrane. When anchored to the lipid part they are known as lipid-anchored proteins. Membrane associated proteins are not really MPs; they are cytoskeletal and associated signaling proteins at the inner leaflet surface and certain glycoproteins at the outer leaflet surface, and are thought to be involved in stabilizing cell membranes. These components are membrane-associated but not involved in the integral structure of membranes and thus they can be removed easily [18]. Integral proteins, on the other hand, are strongly associated to the membrane and thus they require detergents, bile acids, organic solvents or other protein denaturants to dissociate from the membrane. After dissociation from the membrane, however, they remain associated with lipids and therefore are highly insoluble in water [18]. Integral MPs play a key role in detecting and conveying extracellular signals into the cells, allowing them to interact and respond to the environmental signals. Their membrane spanning domain, have the shape of either alfa-helices or beta-barrels. Most of the integral MPs appear to have one or more alfa helical domains, connected by alternating intracellular and extracellular peptide loops conferring a helix conformation. Integral MPs can be divided into two groups: monotopic and polytopic. The monotopic MPs are associated with the membrane from one side but do not span the lipid bilayer completely while polytopic MPs cross the membrane. Polytopic MPs, also known as transmembrane proteins, can be further divided into single-pass and multi-pass MPs. Single-pass MPs cross the membrane once and multi-pass cross the membrane more than one time. Hence, the *transmembrane* proteins interact with the hydrophilic and hydrophobic leaflets of the membrane [19,20].

Generally, biomembranes are modeled as a complex and dynamic spatial-temporal mix of MPs and lipids. The ratio of lipids, proteins as well as lipids to proteins changes constantly. Different asymmetries between the inner and the outer leaflets of the membrane determine the characteristics and function of the membrane itself. It is well established that plasma membrane and organelles' membrane, including endosomes, show differences both in protein and lipid compositions [21–24].

1.2 Membrane proteins in extracellular vesicles

Extracellular vesicles (EVs) are a heterogeneous group of lipid bilayer bound vesicles comprising exosomes, microvesicles and apoptotic bodies 30–1000 nm in diameter and are secreted by all cells. EVs are known to be involved in cellular transport, secretion and intercellular communication allowing cells to exchange proteins, lipids and genetic material [25]. Recent progress in EV research sheds light on the importance of these structures also in interspecies communication [26–28]. EVs can be isolated from various human biological samples, like body fluids [29–32], cell lines [33] and even complex tissues [34]. The isolation of EVs is usually followed by the analysis of their molecularcomposition, which led to the accumulation of a huge amount of data on protein, lipid and RNA cargo contents of EVs. The data obtained in the rapidly growing number of EV studies have been deposited in different publicly available databases, like Vesiclepedia [35], ExoCarta [36] and EVPedia [37]. While, there are a number of excellent reviews that discuss the biocargo composition of EVs, membrane protein component of the EV proteome has been less studied [25,38,39]. EVs have been shown to contain certain important functional classes of MPs (Fig. 1) among these are tetraspanins, lipid anchored proteins, enzymes, transporters, receptors and adhesion molecules that are reviewed here.

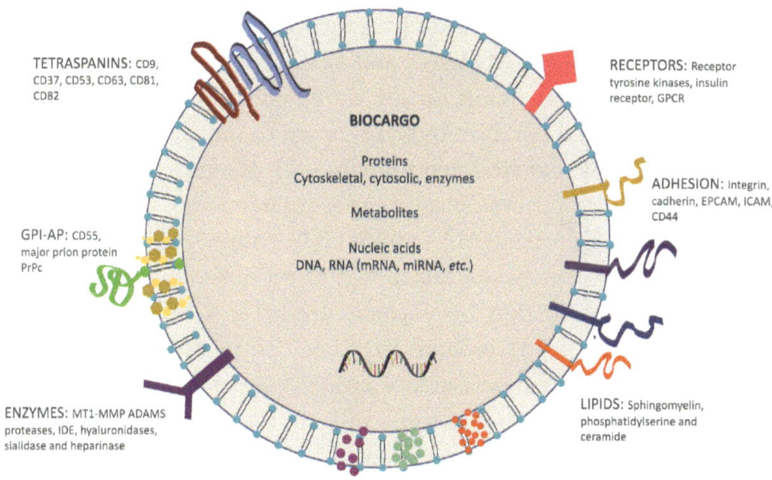

Fig. 1 The biomembrane of extracellular vesicles. The most studied protein and lipid classes are illustrated with examples. *ADAMS, disintegrin and metalloproteinase family proteins; EPCAM, epithelial cell adhesion molecule; GPCR, G protein coupled receptor; ICAM, intercellular adhesion molecule; IDE, insulin degrading enzyme; MT1-MMP, membrane type 1 metalloproteinase.*

1.2.1 Tetraspanins

The tetraspanin family is composed of 33 members in humans and includes the frequently used EV protein markers CD9, CD37, CD53, CD63, CD81 and CD82 [40]. Tetraspanins are known to form membrane microdomains also known as tetraspanins enriched microdomains (TEMs). TEMs control the distribution and grouping of a large variety of proteins and can associate with lipids like cholesterol. In TEMs tetraspanins form clusters and interact with a large variety of cytosolic, extracellular and transmembrane proteins, mostly integrins, immunoglobulin superfamily receptors and metalloproteinases. The membrane compartmentalization of tetraspanins promotes a wide range of intracellular signaling events through the formation of protein signaling complexes, cell trafficking, membrane recognition accompanied by morphological changes such as membrane fusion and fission, fundamental immune cell functions, including adhesion, pathogen uptake and proliferation. Regarding EVs, TEMs were shown to play a role in the cargo loading and have also been implicated in the ESCRT-independent biogenesis of EVs [39]. This was demonstrated when cells depleted of the ESCRT proteins were able to produce CD63 positive EVs. Also, structural analysis of CD81 reveals a cone like structure with an intramembrane pocket that can accommodate cholesterol [25] suggesting that the clustering of these units into a microdomain could induce an inward budding. Moreover, CD81 and CD9 were shown to be involved in cargo sorting to EVs, for example, the genetic suppression of CD81 impairs the inclusion in EVs of the proteins that interact with CD81 (ICAM-1 and EWI-2) [41]. Several tetraspanins were found to be post-translationally modified by palmitoylation in the TEMs. Palmytoylated tetraspanins were shown to interact with other tetraspanins in the TEM and asociated proteins [42]. Romancino et al. pointed out the importance of palmitoylation in the interaction of programmed cell death 6-interacting protein (ALIX) with CD9 within TEMs [43].

Tetraspanins in EVs have been shown to modulate cancer progression: the down or upregulation of Tspan8, Tspan12, CD9, CD37, CD63, CD81, CD82 and CD151 proteins have been correlated with either good or bad prognosis in different types of cancers [44–46]. Zaborowsky et al. applied an *in-silico* approach to identify antigens that can be used to differentiate between cancer- and normal cell-derived EVs [47]. The approach relies on the assumption that antigens presented on the surface of cancer cells are also expressed on the surface of their EVs, providing the ability to isolate them selectively. This, in fact, has been confirmed for a few proteins, like CD47, CD71 and EpCAM for ovarian-cancer [47], CD 147 for colorectal cancer [48], and glypican-1 for pancreatic cancer [49].

1.2.2 Lipid anchored proteins

Lipid anchored proteins and especially glycosylphosphatidylinositol-anchored proteins (GPI-APs) have received much attention in EV research. GPI-APs are composed of a phosphatidylinositol group linked to a carbohydrate that bridges to the C terminal of a protein, hence the GPI moiety allows that these proteins interact with different molecules to perform a wide range of molecular functions. GPI-AP are exclusively localized on the outer leaflet of the bilayer and are preferentially embedded in lipid rafts or detergent resistant membranes (DRM) [50]. DRM regions enriched in GPI-APs have been shown to be preferentially endocytosed to the recycling endosomal compartment and upon maturation form intraluminal vesicles (ILVs) that are then released as EVs. These processes link the GPI-APs to the biogenesis of EVs [25,51].

1.2.3 Enzymes and transporters

Recently several functionally active enzymes were shown to localize on the surface of EVs including metalloproteinase family enzymes (like MT1-MMP), ADAMS proteases, insulin-degrading enzyme (IDE), hyaluronidases, sialidase (like NEU3) and heparanase [52–54]. Besides their functional role to catalyze a diverse arsenal of biochemical reactions, to date, most of the surface enzymes identified in EVs have been linked to diseases. Of particular interest is the insulin IDE that can cleave amyloid β-protein and the β amyloid precursor protein, eliminating the neurotoxic effects of Aβ peptides related to the progression of Alzheimer's disease [52]. Moreover, based on specific enzymatic properties, EVs have been proposed to be employed in tissue repair, especially due to degradation of extracellular matrix and mobilization of growth factors [55].

Transporters participate in the movement of substances across the cell's membranes by utilizing electrochemical gradients. Similarly, to the cell membrane, EVs contain a high percentage of membrane transporters. The presence of aquaporins, for example in EVs isolated from human urine is well studied and has been described in various vesicle types and diseases [56].

1.2.4 Receptors and ligands

A receptor is a protein that receives signals by binding specific ligands, this receptor-ligand mediated interaction at the surface of EVs is required for intercellular communication and cellular uptake processes. In a first step, EVs need docking at the plasma membrane, followed by the activation of surface receptors with their specific ligand protein. EVs have been shown to present both receptors and ligands, and this aspect has been studied especially for tumor and immune cell derived EVs [57,58]. Moreover, EVs can transmit information to acceptor cells by acting at the cell surface, without delivery of their content, for example, EVs derived from dendritic cells carry in their surface MHC class I and II complexes, they can bind to the surface of the T lymphocytes by interacting with the CD4 receptor protein and transmit the information [59].

1.2.5 Adhesion proteins

Adhesion proteins build up structures between and within cells and mediate adhesion between the cell and the surroundings. Cell-cell adhesion is mediated through membrane proteins called cell-adhesion molecules (CAMs). CAMs are further divided into four main families: cadherins, immuno-globulin superfamily (IgS), integrins, and selectins. CAMs are composed of three conserved domains: a cytosolic domain that acts as a linker that connects directly or indirectly with the cytoskeleton, the extracellular domain that acts as adhesive interaction between CAMs of the same type or to a different class of CAMs, and the transmembrane domain [20]. The presence of CAM in EVs has been studied and its expression has been related to different pathologies [53]. Of special interest is the role of EV carrying integrins in cancer diagnosis. Hoshino et al. found that integrins profile in EVs is correlated with certain types of cancer; they found that integrin (ITG) $\alpha_6\beta_4$ and ITG $\alpha_6\beta_1$ were related to lung metastasis whereas ITG$\alpha_v\beta_5$ was associated to liver metastasis. Moreover, metastatic cell EVs contained ITG$\alpha_2\beta_1$ [60]. Altogether, their findings suggest the potential use of EVs for the prediction of tumor progression.

Here, we applied a bioinformatics workflow including functional annotation and GO term enrichment analysis to study the human EV proteome and to identify and classify MPs. We created a dataset called human EV membranome by extracting the MPs from the EV proteome. This was compared to the previously described human membranome with the aim to extract and study proteins of particular functional classes known to be involved in EV bioactivity or related to the biogenesis of EVs.

2. Methods

2.1 Databases and datasets

The human EV dataset was downloaded on October 15, 2019, from publicly available EVpedia database [37]. EVpedia contains 74,696 proteins from 97 studies for eukaryotes of which 25,118 are from the human dataset, containing proteins identified in different EV isolates. From this dataset, a subset of 4,332 proteins identified with a frequency score higher than 20 were extracted for further analysis. FASTA file containing these protein sequences was retrieved using the Batch Retrieve/ID mapping tool of UniProt. The 4,332 human EV protein entries were mapped to 4,283 UniProtKB IDs. In this work, we refer to this set of proteins as Human EV proteome.

Human membrane and surface proteins were downloaded from the recently published *in silico* human surfaceome database (http://wlab.ethz.ch/surfaceome/) that contains information on 20,194 proteins including 5,102 membrane and 2,886 surface proteins [15].

UniProt is a compendium of databases for protein sequence and annotation data [9]. UniProt Knowledgebase is the central hub for the collection of functional information on proteins and consists of 2 sections: SwissProt, manually reviewed with information extracted from literature and curator-evaluated analysis, and TrEMBL that are unreviewed automatically annotated proteins. The *Homo sapiens* proteome was downloaded on November 5, 2019, from UniProt database and contained information of 24 genes that encode for 74,788 proteins of which 20,605 were reviewed and 54,183 were unreviewed. This was used as a reference database in BLAST to query the EV dataset and to make a further functional annotation.

2.2 Bioinformatics tools

Functional annotation and data mining were perform using Blast2GO version 5.2.5 software. Blast2GO (B2G) is a bioinformatics tool for the automatic functional annotation of DNA and protein sequence data, mainly based on the Gene Ontology (GO) vocabulary, it includes numerous functions for the visualization, management, and statistical analysis of annotation results, supports InterPro, Enzyme Codes (EC), KEGG pathways, GO direct acyclic graphs (DAGs), and GOSlim [61]. The Human EV proteome dataset (4,283 proteins) was blasted against the Human Proteome database (20,605 proteins). The Blastp search algorithm was used via local database from *Homo sapiens* UniProt Knowledge database using a number of blasts hits set to 20, an expectation value (e-value) of 1.0E-3 and a minimal alignment length (hsp) of 33. The Blast hits of each protein sequence were mapped with GO terms deposited in the GO database [62]. Annotation was performed setting the parameters as follows Annotation Cut Off to 55, the GO weight to 5, E-value Hit Filter to 1.0 E-6 and a Hit Filter of 500.

InterPro functional analysis search was performed to identify signature sequences in CDD, HAMAP, HMMPanther, HMMpfam, HMMPIR, Fprintscan, BlastproDom databases, all sequences generated InterPro results. InterPro is a resource that provides functional analysis of protein sequences by classifying them into families and predicting the presence of domains and important sites known as signatures, provided by several different databases [63]. InterPro annotation was merged to Blast2GO annotation. All 4,283 sequences were mapped and annotated. The total number of annotations was 25,482. GO direct acyclic graphs (DAG) were created using Blast2Go.

FunRich, standalone software version 3.1.3 was used for comparison of proteins from different datasets [64].

QuickGO version 2.0 was used to study the enrichment of diverse binding activities (carbohydrate, protein, lipid and nucleic acid) annotations in the EV proteome compared to GO annotations in the human proteome. QuickGO is a web-based tool that allows browsing of the gene ontology and all associated annotations, it can be queried for multiple proteins to analyze the distribution of GO terms [65].

WebGesalt web server was used to perform a gene over representation analysis (ORA) to find out if the differentially expressed gene products are associated with a certain biological process, molecular function or cellular component; it is done by comparing the frequency of individual annotations in a gene list with a reference list [66]. The parameters were set as follows: minimum number of genes for a category was set to 5, the maximum number of genes for a category was set to 2000, the multiple test adjustment was set to the BH method (Benjamin and Hochberg), the significance level for FDR was set to < 00.05, the top significance Level was set to 10, and the number of categories expected from set cover was set to 10.

3. Results

3.1 The human EV proteome *in silico*

In this work, we have used the EVPedia database for the extraction of EV proteins because it provides taxonomy ID and frequency scores. Frequency score is a number that shows how many times a protein or its ortholog has been identified. Human proteins (Human, TaxID: 9606) with at least 20 frequency score were extracted from EVPedia database (4,332 sequences, Supplementary Table S1). Functional annotation and GO term enrichment analysis led to 25,482 GO annotations along the three ontologies: biological process (BP), molecular function (MF) and cellular component (CC) (4283 sequences, Supplementary Table S2). It should be noted that the protein evidence in individual GO terms is not mutually exclusive as the presence of one given protein may be reflected in several of the GO ontologies. 85, 72 and 51% of the total protein sequences were annotated with MF, BP and CC categories, respectively. These were separately analyzed to reveal characteristic EV related features.

3.1.1 BP GO term annotation analysis

Second level BP GO categories (3,118 sequences, 100%) were distributed in 9 different GO terms. The main BP GO categories were cellular process (GO:0009987, 76%) followed by metabolic process (GO:0008152, 58%) and biological regulation (GO:0065007, 40%), but response to stimulus (GO:0050896), signaling (GO:0023052) and establishment of localization (GO:0051179) were highly expressed as well. This BP distribution was found to be similar to that previously obtained for the proteins of EVs isolated from blood [29] and urine [30,67]. To obtain more specific EV related data we have extracted and analyzed the distribution of two selected BP GO terms: transport (GO:0006810) and cell communication (GO:0007154).

Transport (GO:0006810) term was selected because transport of macromolecules, small molecules and ions into and out of an EV is a very important process related to the dynamic character of the vesicles. Indeed, we have found a high number of proteins associated with BP GO term transport (795 sequences highlighted in blue in Supplementary Table S2). These are mainly related to ion transport (220 proteins, GO:0006811), vesicle mediated transport (354 proteins, GO:0016192), intracellular transport (208 sequences, GO:0006886) and nitrogen compound transport (322 proteins, GO: 0071705), the latter was found to be highly enriched in protein transport related proteins (280 sequences). Intracellular transport GO terms in our data set are mainly associated with proteins targeting to the mitochondrion, nucleus and cell membrane so could not truly represent EV proteins. On the other hand, vesicle-mediated transport GO terms were found to be enriched in exocytosis, endocytosis and phagocytosis, and endosomal transport pathways related proteins. This latter contains proteins related to the retrograde transport from endosome back to the trans Golgi (21 proteins), but importantly also proteins that support the endosomal origin of certain types of EVs like endosomal transport via the multivesicular bodies (MVBs) (8 proteins), retrograde transport to the plasma membrane (4 proteins) and endocytic recycling related proteins (15 proteins) GO terms.

A very high number of sequences were associated to BP GO term "cell communication" (GO:0007154) which is in line with our current understanding of the role of EVs in intercellular communication. Most of them are associated to "signal transduction", a process in which a signal triggers the activity of a cell, for example by receptor activation or ligand binding. More importantly, we have found 267 surface receptors, out of which 80 are from the transmembrane receptor tyrosine kinases (RTKs) signaling pathways which are involved in epidermal growth factor receptor (EGFR) signaling (18 sequences), insulin receptor (IR) signaling (8 sequences), platelet-derived (9 sequences) and fibroblast growth factor receptor (FGFR) binding (12 sequences) (Supplementary Table S3). The ErbB family of proteins contains four receptor tyrosine kinases, structurally related to the EGFR receptors signal through Akt, MAPK, and many other pathways that are known to regulate cell proliferation, migration, differentiation, apoptosis, and cell motility and have important roles in cancer. Indeed, previous studies documented the presence of EGFR in the cargo of tumor-derived EVs [68]. In addition, recent data shows the active participation of EVs in the pathogenesis of type 2 diabetes mellitus and thus in insulin receptor signaling [69]. EV-associated surface receptors could be an important set of potential biomarkers and therapeutic targets for various pathologies including tumor (FGFR, EGRF) and diabetes (IR). Using *in silico* analysis and data mining approach here we show that a relevant set of proteins, like those related to the RTKs pathways, can be extracted and evaluated for EV-based targeted therapy approaches.

3.1.2 MF GO term annotation analysis

MF GO analysis at the second level (3,670 sequences, 100%) shows a distinct distribution of only two categories: binding (GO:0005488, 63%) and catalytic activity (GO:0003824, 36%). The peculiarity of the high expression of "binding" GO term in the human EV proteome might be explained by the fact that EVs in order to transfer their cargo to the recipient cell necessarily need to bind to different molecules. To analyze this aspect we have extracted from the MF GO DAG those terms that were related to (i) nucleic acid binding (GO:0003676), (ii) protein binding (GO:0005515), (iii) carbohydrate binding (GO:0030246) and (iv) lipid binding (GO:0008289). These

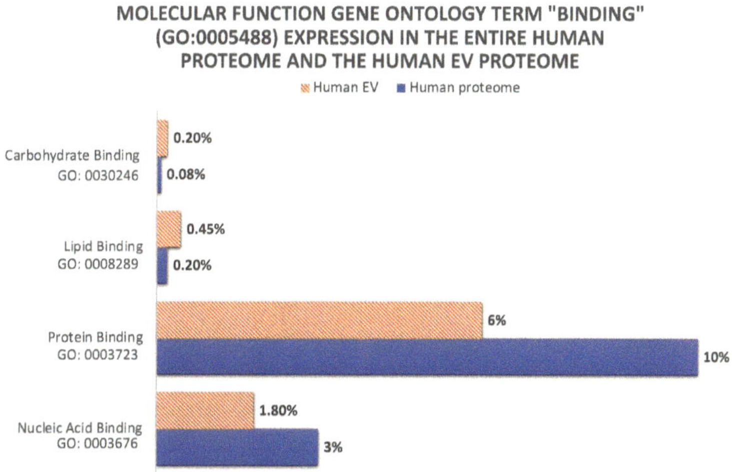

Fig. 2 Comparison of selected binding related child GO terms between the entire human and the EV proteomes. The human EV proteome related data is displayed in orange, Human proteome is displayed in blue. *QuickGo was used to display entire human proteome related GO term data [65].

categories were selected because they are the principal molecular classes that are present on the surface of recipient cells (proteins, lipids and carbohydrates) and are of particular interest to design the directed uptake of the vesicles. In addition, these specific binding properties can be exploited for the isolation of specific EV subclasses. Fig. 2 shows the comparison of selected binding related child GO terms between the entire human and the EV proteomes. This analysis highlights that while protein and nucleic acid bindings are less enriched in the EVs, carbohydrate and lipid bindings are more prominent in EVs when compared to the entire human proteome. Looking deeper into this data, we found 482 proteins related to nucleic acid binding (GO:0003676) with a high representation of RNA binding (GO:0003723, 203 sequences). This is in line with the EV-mediated horizontal transfer of RNA and regulation of gene expression in recipient cells that have been extensively reported in recent years [70]. In the EV dataset, there are 36 proteins related to translation factor activity, 20 to mRNA binding, 7 to tRNA binding and 7 to rRNA binding, highlighting the diversity of different RNAs that are present in EVs, and these are highlighted in green in Supplementary Table S2. Indeed different studies have demonstrated the presence of various types of RNA in EVs, like rRNA and tRNA, miRNA, vault RNA and Y RNA. This is in accordance with the richness of RNA binding proteins found in EVs [71,72]. RNA binding proteins participate in the packaging of RNA into the EVs but also in the immune regulation through the exchange of different RNAs from one cell to another, the later one is of great interest for their use as a potential biomarker in different diseases [73].

A high number of proteins with protein binding GO term (GO: 0005515) was found in the human EV proteome dataset (1,572 sequences), mainly enzymes (GO: 0019899), cytoskeletal proteins (GO:0008092) and signaling receptor (GO: 0005102) binding proteins. The signaling receptor binding proteins in EVs (178 sequences) can selectively and non-covalently interact with receptors and can be particularly important for the specific cellular uptake of EVs. In this category, there are the G-protein-coupled receptor (GPCRs) binding proteins (GO:0001664, 25 sequences, highlighted in yellow in Supplementary Table S2) that have been identified in different EVs isolates, including human saliva [74]. We have found only a relatively low number of carbohydrate binding proteins (GO:0030246) in EVs (53 sequences) but it is worth to note they are enriched in comparison to the entire human proteome (Fig. 2). Among these, a higher number of proteins show an affinity for monosaccharides, mainly ascorbic acid and mannose, but a few polysaccharides binding proteins were also identified in human EV isolates. We have found 116 lipid binding proteins (GO:0008289) out of which 74 were phospholipid binding proteins (GO:0005543). The lipid binding proteins have been somehow overlooked in EV research. However, in the past few years, they attracted more attention due to their potential for the development of affinity-based EV isolation methods [75]. Furthermore, lipids are important in the biogenesis, uptake and functions of EVs [76]. Finally, knowledge of protein, carbohydrate and lipid binding proteins in the EV proteome can be particularly important when targeted cellular uptake of EVs is the aim.

3.1.3 CC GO term annotation analysis

Regarding the CC, the second level GO categories of CC (2225 sequences, 100%) showed a distribution in 3 main groups: intracellular (GO:0005622, 35%), cellular anatomical entity (GO:0110165, 49%) and protein-containing complex (GO:0032991, 21%). The cellular anatomical entity is described in the next section. Here we further analyze only the intracellular category. A closer look at them shows 825 cytoplasmic proteins (GO: 0005737) and 1,224 proteins related to intracellular organelles (GO: 0043229). The set of cytoplasmic proteins identified here could be the soluble component of the vesicle proteome, while the high number of intracellular organelles related proteins could be explained by the shared components between intracellular vesicles, like the intraluminal vesicles (ILV) of the MVBs and EVs. However, we cannot exclude that partially they could indicate the intracellular vesicle contamination in the human EV protein data deposited in EVPedia.

3.2 EV membrane proteins

EV MPs have fundamental roles in the biogenesis as well as in other biological processes such as immunomodulation, cancer progression, cellular communication, tissue regeneration. To better understand these roles, we have analyzed the MPs data set of the human EV proteome deposited in EVPedia. Two methods were used to define EV MPs: (i) CC GO term analysis of the human EV proteome dataset and (ii) creating a reference data set using the union of the EV proteome and human membranome databases [15]. Merging the two dataset ensued a group of 1,299 proteins that we called "human EV membranome" (Fig. 3).

3.2.1 EV membrane proteins based on the CC GO term analysis

By looking at CC GO term of the EV proteome at annotation level 3 we have found a high number of proteins (1,009 sequences, 45.3%) with the "membrane" GO term (GO: 0016020) (Fig. 4A). The membrane GO term refers to all the proteins and protein complexes that are embedded or attached to a lipid bilayer. The higher percentage of membrane proteins in the human EV proteome in comparison with the human proteome was expected because of the increased surface area to volume ratio with

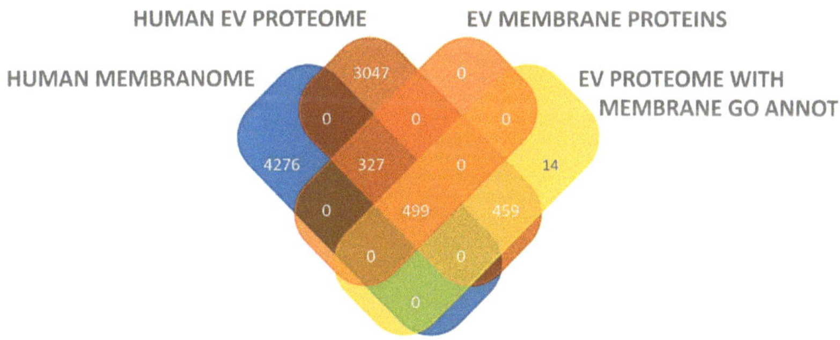

Fig. 3 Venn diagram of the different datasets used in this work that contains the human membranome [15] the human EV proteome extracted from the EVPedia [77] with frequency score higher than 20. The EV membranome contains all the membrane proteins extracted from the human EV proteome database using cellular component GO term analysis and EV proteins present in the human membranome database.

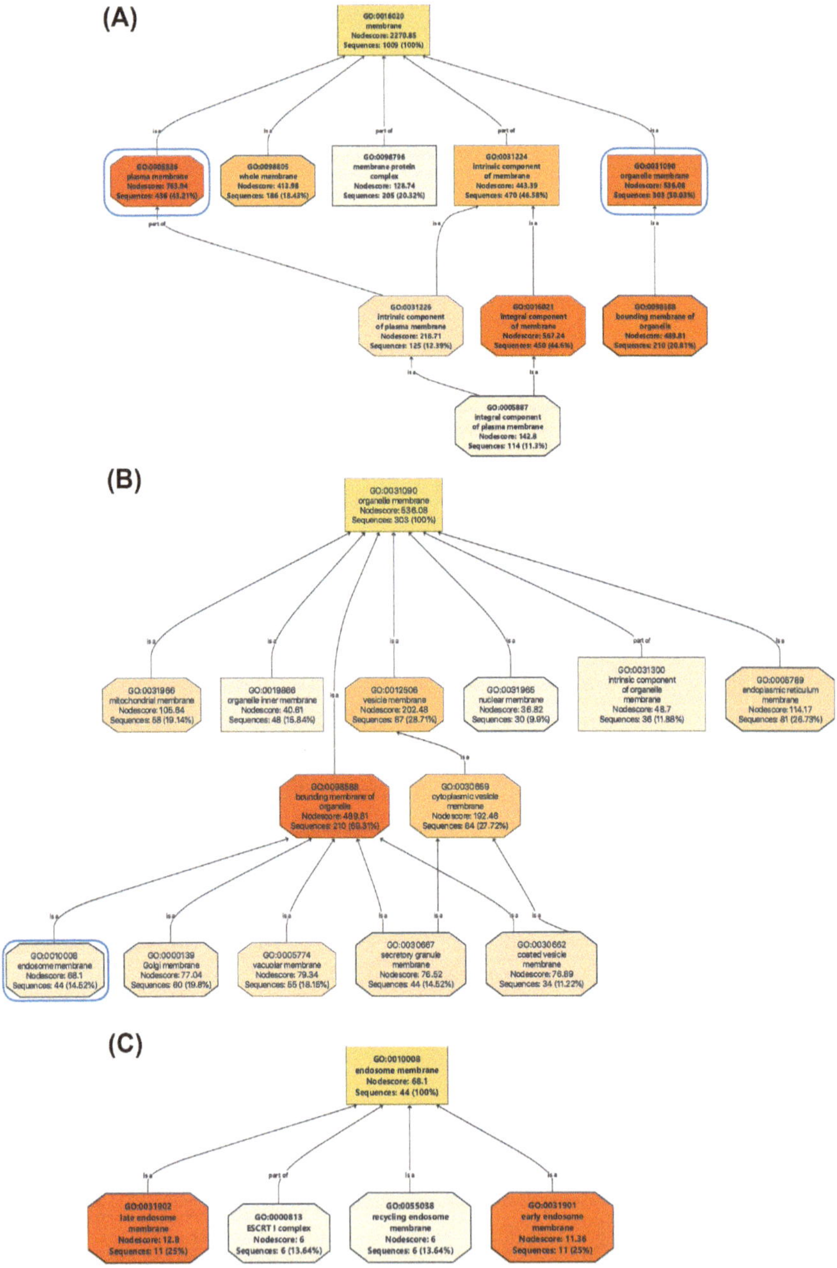

Fig. 4 Gene ontology (GO) direct acyclic graphs for (A) Membrane (GO:0016020, (B) Organelle membrane (GO:0031090) and (C) Endosome membrane (GO:0010008) GO associated terms. Each DAG displays the parent and children nodes with its respective node score and the number of protein sequences in the node.

decreasing internal diameter. In our dataset, we have found two highly represented membrane GO child terms: plasma membrane (GO:0005886, 436 sequences) and organelle membrane (GO003109, 303 sequences). One would anticipate that vesicles formed by outward budding are enriched in MPs originated from the plasma membrane (GO:0005886), while the late-endosome or MVBs derived EVs are enriched in MPs surrounding the endosome (organelle) membrane (GO:0010008). The endosome membrane is formed from the plasma membrane by two consecutive invaginations but during this process the membrane composition, which is highly dynamic, changes considerably. To confirm this, we have analyzed the organelle membrane GO term (GO:0031090) DAG (Fig. 4B). Fig. 4B shows that the node scores of child terms of "organelle membrane" are quite evenly distributed among mitochondrial (58 sequences), endoplasmic reticulum (81 sequences), nuclear (30 sequences), vacuolar (55 sequences), Golgi (60 sequences), cytoplasmic vesicles (84 sequences), and endosome membrane (44 sequences) and none of these "organelle membranes" are particularly enriched in our dataset. Consequently, the endosome membrane is not highly represented in our data. The presence of numerous different intracellular organelles related MPs in human EV samples might also indicate a possible co-purification of these organelles with EVs. Cell damage followed by cell lysis in cultured cell lines is often understated because the cell will no longer be detected by cell counting measurement devices. A recent study showed that lysed cells can count for a significant portion (up to 11%) of the cultured cells, so the EV community should be aware of that when interpreting molecular and functional data [78]. Therefore, it is expected that intracellular organelles and vesicles are present in the culture media and can co-purify with EVs. The low percentage of proteins associated with endosome membrane GO term (GO:0010008, 44 proteins) in the human EV proteome could indicate the low level of EVs with an endosomal origin. Although these proteins (Fig. 4C), like the different components of the Endosomal Sorting Complexes Required for Transport (ESCRTs, 11 proteins) and late endosome membrane proteins (11 proteins) are important and frequently studied in EVs.

3.2.2 EV membrane proteins: the intersect of EV proteome and human membranome databases

As an alternative method to the CC GO term analysis for the extraction of membrane proteins, we compared the frequently identified proteins in the human EV proteome database (4,332 proteins in Supplementary Table S1) with the complete human Membranome (5,102). This comparison resulted in 826 common proteins referred here as "EV MPs" (Fig. 3). The protein list (Supplementary Table S4 part B) was used for functional classification, topology determination and subcellular localization.

3.2.2.1 Functional classification

Functional analysis of the group EV MPs was performed according to the classification described by Almen et al. [5]. Five functional classes, i.e. enzymes, receptors, transporters, miscellaneous and unclassified were considered and their relative distributions were calculated in human membranome and EV MPs. The results of this comparative analysis are shown in Fig. 5. The comparison of the human membranome with EV MPs reveals similarities in the distribution of the enzymes and transporters functional classes and shows a high number of unclassified proteins. Although, significant differences were found in the percentages of receptors and miscellaneous categories. Miscellaneous proteins among others include ligands and structural adhesion proteins. This is interesting because ligands are known to have an important role in EV-cell "cross-talking". Phospholipid bilayer contains, among other proteins, receptors and ligands. Receptors as recognition elements are expected to be more abundant on the recipient cell surface, while the ligands are more abundant on the EV surface. Structural adhesion proteins, on the other hand, can build up structures

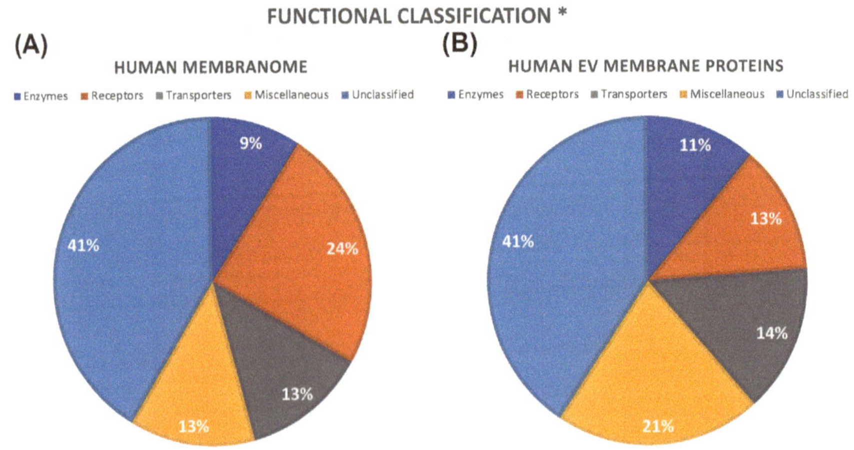

Fig. 5 Functional classification of membrane proteins according to Almen et al. [5]. (A) The human membranome database composed by 5,102 sequences and (B) The human EV membrane proteins set is composed by 826 proteins.

between or within cells or mediate adhesion between the cell and the surroundings. Indeed it has been demonstrated that EVs form aggregates and easily attach to the surfaces [79]. Moreover, it is also a property that helps for the internalization of EVs into the cells. Therefore, the enrichment of the miscellaneous and the decrease of receptor protein functional classes in EV MPs respect to the cell membranome seems to be in line with the specific functional characteristics of EVs.

3.2.2.2 Topology classification

MPs can be classified according to their topology, which divides the proteins in the following categories: singles-pass MP, multi-pass MP, lipid anchored MP and peripheral MP. Comparison of the EVs MPs and the human membranome (Fig. 6) shows little difference in the group of lipid anchored proteins and the peripheral proteins while single-pass proteins and multi-pass proteins indicate considerably differences between the two datasets.

The relative percentage of the single-pass MP group was higher for the EV MPs. They represent the most abundant and functionally diverse category of MPs. These proteins are composed of a transmembrane domain, an intracellular domain and an extracellular domain. Usually the extracellular domain is the longest one, sometimes even 6 times longer than the intracellular domain. Single-pass MPs thus interact with a wide array of

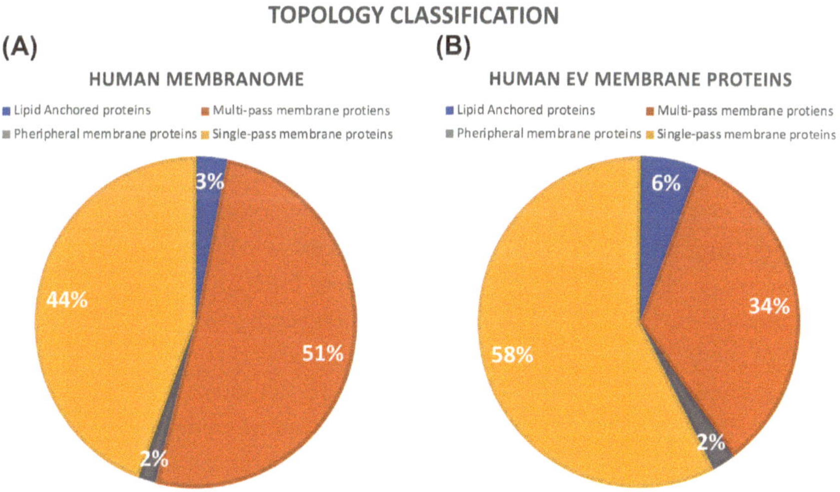

Fig. 6 Topology classification of membrane proteins of the human membranome and the EV membrane proteins. (A) The human membranome database composed by 5,102 proteins, of (B) The human EV membrane proteins composed by 826 proteins.

substrates [80]. The enrichment of single pass MPs in EVs may have implication on their abilities to interact with a wide array of proteins on the target cell, to transfer the biocargo and to fulfill their function as messengers.

3.2.2.2.1 GPCRs in the human EV-MPs Many multi-pass MPs are transporters and receptors, including the GPCRs which are characterized by 7-TM domains. These proteins are of special interest since they are the targets of 40% of all commercialized drugs in the pharmaceutical industry. Our data show a high percentage of these proteins in the human membranome but less in EVs (only 2% of the EV MPs have 7 TM domains). These indicate that the receptor function in EVs could be less pronounced than in the cell membrane.

3.2.2.2.2 Tetraspanins in the human EV-MPs Another group of proteins that belongs to the multi-pass category is the tetraspanins, which are scaffolding proteins characterized by a 4-TM domain. Among the frequently identified EV proteins, we could find 18 out of 33 tetraspanins, and these are listed in Supplementary Table S5. These includes 11 TSN proteins (TSN1, TSN3, TSN4, TSN5, TSN6, TSN7, TSN8, TSN9, TSN11, TSN14, TSN15), 6 CD proteins (CD9, CD53, CD63, CD81, CD82 and CD151) and uroplakin 1a (UPK1A). Tetraspanins are often used as specific protein biomarkers of EVs. For example, EVs enriched in late endosome components are supposed to express CD63. On the other hand, CD9 and CD81 are mainly located in the plasma membrane and therefore they are expected to be implicated in the formation of EVs through the membrane budding mechanisms. Although this is long been a subject of debate [40,58,81].

3.2.2.2.3 GPI-anchored proteins in the human EV-MPs 40 GPI-APs were found by filtering the MPs of EV using the "surfaceome" label (Supplementary Table S6). GPI APs are known to have a wide range of functions and are frequent candidate biomarkers for different diseases. In a recent work, Kooijmans et al. have created GPI-anchored anti-EGFR nanobodies on EVs to promote tumor cell targeting [51]. In our dataset, most of the GPI-APs belonged to the miscellaneous category (17 proteins). One example is the major prion protein PrPc (Supplementary Table S6) known to be involved in diverse processes including neuronal differentiation, cell adhesion, signal transduction, synaptic function and copper homeostasis. EVs decorated with PrPc demonstrate a neuroprotective role. It is interesting because misfolded PrPc is involved in prion diseases, a group of neurodegenerative diseases characterized by the degeneration of the brain and their transmissibility among species [82]. Recently, EVs have been proposed as a source for the intercellular spreading of misfolded prion proteins PrPc. Similarly, the role of EVs in the spreading of Alzheimer's and Parkinson's disease has also been put forward [83]. We have identified 5 GPI anchored enzymes frequently identified in human urinary EVs, like Prostate stem cell antigen, Xaa-Pro aminopeptidase 2, Semaphorin-7A, Glypican-4, Pantetheinase. Moreover, 4 GPI anchored receptors and 3 transporters have been also individuated (Supplementary Table S6).

3.2.3 Enrichment analysis of the EV membranome

Enrichment analysis of the EV membranome (Fig. 3) was performed against the human membranome as a background database [15] to show highly expressed EV MPs and to underline EV related processes. Results of the enrichment analysis are shown in Fig. 7.

In the BP section (Fig. 7A), 4 processes were found to be enriched and these were all related to immunity: the myeloid cell activation involved in immune response (GO:0002275), cell activation involved in immune response (GO:0002263), myeloid leukocyte activation (GO:0002274) and immune effector process (GO:0002252). These results underline the importance of EV in immunity and support the findings of several works that show how EVs trigger the immune response both in healthy and pathologic states [84,85]. We also found that the EV membranome was enriched in different other processes, like exocytosis (GO:0006887), vesicle mediated transport (GO:0016192) and establishment of localization (GO:0051234) (Fig. 7A). Exocytosis related GO term (GO:0006887) enrichment promotes exocytosis as an alternative way of EV secretion. Several proteins and lipids participate in this process. Initially, Rab GTPases tether and dock the MVB to the plasma membrane, followed by the MVB fusion with the plasma membrane, and then the GTPase RhoA alters the actin polymerization to induce the budding and the consequent release of EVs. However, it is important to note that other exocytosis pathways also exist [21,68].

The CC ORA results (Fig. 7B) show enrichment for the categories related to the extracellular space, so we could say that the EV membranome dataset is enriched in proteins that are oriented to the extracellular milieu. Highly enriched categories are extracellular space (GO:0005615), extracellular region part (GO:0044421) and extracellular region (GO:0005576) but the secretory granule (GO:0030141) and vesicles (GO:0031982) categories were also enriched.

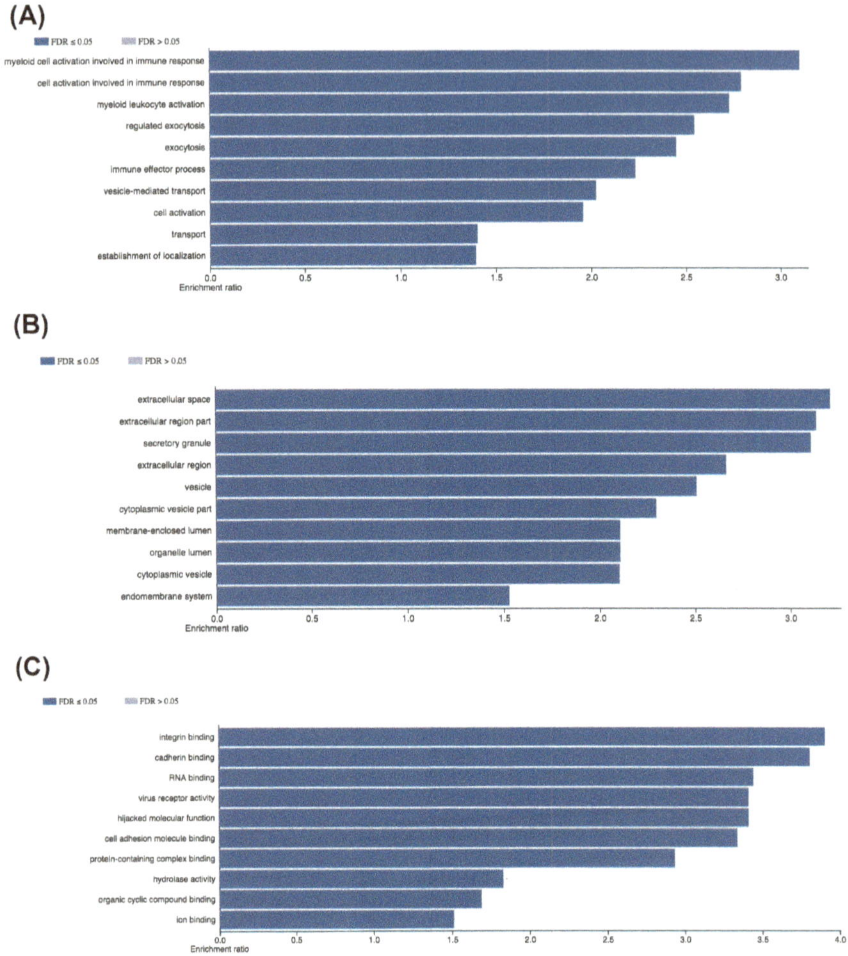

Fig. 7 Results of enrichment analysis based on WebGesalt [66]. The top 10 gene ontology (GO) term annotation results related to (A) Biological Process (BP), (B) Cellular Component (CC), (C) Molecular Function (MF), with their corresponding enrichment ratio from the ORA analysis.

The MF ORA results (Fig. 7C) show an enrichment for the following categories: integrin binding (GO:0005178), cadherin binding (GO: 0045296), RNA binding (GO:0003723), virus receptor activity (GO: 0001618), cell adhesion molecule binding (GO:0050839) and protein-containing complex binding (GO:0044877). These results somehow mirror the results obtained by the B2G analysis of EV proteome in which binding was widely distributed in our dataset. It also highlights the importance of adhesion molecules on EVs surfaces such as cadherins which are related to cell adhesion, recognition, signaling, communication, angiogenesis and morphogenesis. The variety of functions is mediated by the extracellular domain, whereas the intracellular domain associates with a network of cytoskeletal and signaling molecules, usually composed of α-, β-, and γ-catenins [86]. The cadherin−catenin complex plays a key role in cellular adhesion, which is of great importance for EVs since it let them attach to the receptor cell surface. Recently, various studies have shown that abnormalities in cadherin signals are strongly associated with EVs derived from tumors. Epithelial-mesenchymal transition (EMT) occurs at the initial stage of tumor metastasis and is characterized by down-regulation of the E-cadherin and up-regulation of N-cadherin, tumor-derived EVs transfer EMT-related RNAs and proteins to recipient cells, causing the instability of cadherins, and promoting cancer progression and metastasis [87].

4. Conclusion

Current literature lacks comprehensive information about the protein component of EV biomembrane (Fig. 1) including its role in biogenesis, uptake, functions and topology, and no effort has been made to create a repository database. Recent position paper of ISEV pointed out that there is still a substantial opportunity for research in the EV membrane field [81]. To fill this gap, here we created a subproteome database, the human EV membranome (Fig. 3), which could be useful for structural and functional studies. To demonstrate the utility of the database, functional classes, such as tetraspanin network proteins (Supplementary Table S5), GPI-anchored proteins (Supplementary Table S6), GPCR binding proteins (Supplementary Table S7) and transmembrane receptor tyrosine kinases signaling pathways related surface receptors (Supplementary Table S3) datasets were extracted and analysed. *In silico* analysis of the EV membranome shows that MPs play fundamental roles in (i) EV-cell communication by signaling, (ii) EV uptake by exhibition and adhesion of ligand molecules, (iii) establishment of biocargo by active transport and (iv) biogenesis by budding or endocytosis and membrane fusion related processes.

Membrane proteins are of great medical importance, as indicated by the fact that 66% of approved drugs listed in the DrugBank database target MPs. From a clinical perspective, understanding the membrane proteome is essential for developing EVs as biomarkers. Therefore, we believe that data published here Supplementary Table S3, S5, S6 and S7 like, for example the set of frequently identified GPI-APs, could aid the translation of EVs to clinical applications. Importantly, the EV membranome also contains relevant information on proteins that are enabled for future labeling and functionalization strategies thus facilitating their use as drug delivery system.

Supplementary materials

Supplementary data related to this article can be found online at https://doi.org/10.1016/bs.abl.2020.09.002.

Acknowledgments

GP thanks to VES4US FETOPEN project that has received funding from the European Union's Horizon 2020 research and innovation program under grant agreement No 801338.

References

[1] D.M. Engelman, Membranes are more mosaic than fluid, Nature 438 (2005) 578–580, https://doi.org/10.1038/nature04394.
[2] G. Van Meer, D.R. Voelker, G.W. Feigenson, Membrane lipids: where they are and how they behave, Nat. Rev. Mol. Cell Biol. 9 (2008) 112–124, https://doi.org/10.1038/nrm2330.
[3] G. Guigas, M. Weiss, Effects of protein crowding on membrane systems, Biochim. Biophys. Acta Biomembr. 1858 (2016) 2441–2450, https://doi.org/10.1016/j.bbamem.2015.12.021.
[4] E. Wallin, G. Von Heijne, Genome-wide analysis of integral membrane proteins from eubacterial, archaean, and eukaryotic organisms, Protein Sci. 7 (1998) 1029–1038, https://doi.org/10.1002/pro.5560070420.
[5] M.S. Almén, K.J.V. Nordström, R. Fredriksson, H.B. Schiöth, Mapping the human membrane proteome: a majority of the human membrane proteins can be classified according to function and evolutionary origin, BMC Biol. 7 (2009) 50, https://doi.org/10.1186/1741-7007-7-50.
[6] E. London, Biomembranes: molecular structure and function, Robert B. Gennis, Q. Rev. Biol. (1989) 85–103, https://doi.org/10.1086/416484.
[7] D. Bausch-Fluck, E.S. Milani, B. Wollscheid, Surfaceome nanoscale organization and extracellular interaction networks, Curr. Opin. Chem. Biol. 48 (2019) 26–33, https://doi.org/10.1016/j.cbpa.2018.09.020.
[8] H.M. Berman, J. Westbrook, Z. Feng, G. Gilliland, T.N. Bhat, H. Weissig, I.N. Shindyalov, P.E. Bourne, The protein Data Bank, Nucleic Acids Res. 28 (2000) 235–242, https://doi.org/10.1093/nar/28.1.235.
[9] T.U. Consortium, UniProt: a worldwide hub of protein knowledge, Nucleic Acids Res. 47 (2018) D506–D515, https://doi.org/10.1093/nar/gky1049.
[10] M.A. Lomize, I.D. Pogozheva, H. Joo, H.I. Mosberg, A.L. Lomize, OPM database and PPM web server: resources for positioning of proteins in membranes, Nucleic Acids Res. 40 (2011) D370–D376, https://doi.org/10.1093/nar/gkr703.
[11] G.E. Tusnády, L. Kalmár, I. Simon, TOPDB: topology data bank of transmembrane proteins, Nucleic Acids Res. 36 (2007) D234–D239, https://doi.org/10.1093/nar/gkm751.

[12] M.H. Saier, V.S. Reddy, B.V. Tsu, M.S. Ahmed, C. Li, G. Moreno-Hagelsieb, The transporter classification database (TCDB): recent advances, Nucleic Acids Res. 44 (2016) D372—D379, https://doi.org/10.1093/nar/gkv1103.
[13] A.L. Lomize, J.M. Hage, I.D. Pogozheva, Membranome 2.0: database for proteome-wide profiling of bitopic proteins and their dimers, Bioinformatics 34 (2018) 1061—1062, https://doi.org/10.1093/bioinformatics/btx720.
[14] L. Dobson, I. Reményi, G.E. Tusnády, The human transmembrane proteome, Biol. Direct 10 (2015) 31, https://doi.org/10.1186/s13062-015-0061-x.
[15] D. Bausch-Fluck, U. Goldmann, S. Müller, M. van Oostrum, M. Müller, O.T. Schubert, B. Wollscheid, The in silico human surfaceome, Proc. Natl. Acad. Sci. U.S.A. 115 (2018) E10988—E10997, https://doi.org/10.1073/pnas.1808790115.
[16] T. Orioli, M. Vihinen, Benchmarking subcellular localization and variant tolerance predictors on membrane proteins, BMC Genom. 20 (2019) 547, https://doi.org/10.1186/s12864-019-5865-0.
[17] K. Shimizu, W. Cao, G. Saad, M. Shoji, T. Terada, Comparative analysis of membrane protein structure databases, Biochim. Biophys. Acta Biomembr. 1860 (2018) 1077—1091, https://doi.org/10.1016/j.bbamem.2018.01.005.
[18] G.L. Nicolson, The Fluid - mosaic Model of Membrane Structure: still relevant to understanding the structure, function and dynamics of biological membranes after more than 40 years, Biochim. Biophys. Acta Biomembr. 1838 (2014) 1451—1466, https://doi.org/10.1016/j.bbamem.2013.10.019.
[19] R.J. Turner, Understanding the biogenesis of polytopic integral membrane proteins, J. Membr. Biol. 192 (2003) 149—157, https://doi.org/10.1007/s00232-002-1071-z.
[20] L. Harvey, B. Arnold, Z. S Lawrence, l M. Pau, B. David, D. James, Molecular Cell Biology, fourth ed., 2000, https://doi.org/10.1016/j.jasms.2009.08.001.
[21] S.L. Schmid, R. Fuchs, P. Male, I. Mellman, Two distinct subpopulations of endosomes involved in membrane recycling and transport to lysosomes, Cell 52 (1988) 73—83, https://doi.org/10.1016/0092-8674(88)90532-6.
[22] H.J. Sharpe, T.J. Stevens, S. Munro, A comprehensive comparison of transmembrane domains reveals organelle-specific properties, Cell 142 (2010) 158—169, https://doi.org/10.1016/j.cell.2010.05.037.
[23] D. Casares, P.V. Escribá, C.A. Rosselló, Membrane lipid composition: effect on membrane and organelle structure, function and compartmentalization and therapeutic avenues, Int. J. Mol. Sci. 20 (2019) 2167, https://doi.org/10.3390/ijms20092167.
[24] C. Bissig, J. Gruenberg, Lipid sorting and multivesicular endosome biogenesis, Cold Spring Harb. Perspect. Biol. . 5 (2013), https://doi.org/10.1101/cshperspect.a016816.
[25] G. van Niel, G. D'Angelo, G. Raposo, Shedding light on the cell biology of extracellular vesicles, Nat. Rev. Mol. Cell Biol. 19 (2018) 213—228, https://doi.org/10.1038/nrm.2017.125.
[26] J. Mu, X. Zhuang, Q. Wang, H. Jiang, Z.-B. Deng, B. Wang, L. Zhang, S. Kakar, Y. Jun, D. Miller, H.-G. Zhang, Interspecies communication between plant and mouse gut host cells through edible plant derived exosome-like nanoparticles, Mol. Nutr. Food Res. 58 (2014) 1561—1573, https://doi.org/10.1002/mnfr.201300729.
[27] E. Woith, G. Fuhrmann, M.F. Melzig, Molecular sciences extracellular vesicles-connecting kingdoms, Int. J. Mol. Sci. 20 (2019) 5695, https://doi.org/10.3390/ijms20225695.
[28] S. Jimenez-Jimenez, K. Hashimoto, O. Santana, J. Aguirre, K. Kuchitsu, L. Cárdenas, Emerging roles of tetraspanins in plant inter-cellular and inter-kingdom communication, Plant Signal. Behav. 14 (2019) 1—11, https://doi.org/10.1080/15592324.2019.1581559.
[29] M. Rezeli, O. Gidlöf, M. Evander, P. Bryl-Górecka, R. Sathanoori, P. Gilje, K. Pawłowski, P. Horvatovich, D. Erlinge, G. Marko-Varga, T. Laurell, Comparative

proteomic analysis of extracellular vesicles isolated by acoustic trapping or differential centrifugation, Anal. Chem. 88 (2016) 8577–8586, https://doi.org/10.1021/acs.analchem.6b01694.
[30] L.A. Erozenci, F. Böttger, I.V. Bijnsdorp, C.R. Jimenez, Urinary exosomal proteins as (pan-)cancer biomarkers: insights from the proteome, FEBS Lett. 593 (2019) 1580–1597, https://doi.org/10.1002/1873-3468.13487.
[31] A. Cvjetkovic, S.C. Jang, B. Konečná, J.L. Höög, C. Sihlbom, C. Lässer, J. Lötvall, Detailed analysis of protein topology of extracellular vesicles-evidence of unconventional membrane protein orientation, Sci. Rep. 6 (2016) 1–12, https://doi.org/10.1038/srep36338.
[32] C. de la Torre Gomez, R. V Goreham, J.J. Bech Serra, T. Nann, M. Kussmann, "Exosomics"—a review of biophysics, biology and biochemistry of exosomes with a focus on human breast milk, Front. Genet. 9 (2018) 92. https://www.frontiersin.org/article/10.3389/fgene.2018.00092.
[33] R.A. Haraszti, M.-C. Didiot, E. Sapp, J. Leszyk, S.A. Shaffer, H.E. Rockwell, F. Gao, N.R. Narain, M. DiFiglia, M.A. Kiebish, N. Aronin, A. Khvorova, High-resolution proteomic and lipidomic analysis of exosomes and microvesicles from different cell sources, J. Extracell. Vesicles 5 (2016) 32570, https://doi.org/10.3402/jev.v5.32570.
[34] L.J. Vella, B.J. Scicluna, L. Cheng, E.G. Bawden, C.L. Masters, C.-S. Ang, N. Willamson, C. McLean, K.J. Barnham, A.F. Hill, A rigorous method to enrich for exosomes from brain tissue, J. Extracell. Vesicles 6 (2017) 1348885, https://doi.org/10.1080/20013078.2017.1348885.
[35] M. Pathan, P. Fonseka, S. V Chitti, T. Kang, R. Sanwlani, J. Van Deun, A. Hendrix, S. Mathivanan, Vesiclepedia 2019: a compendium of RNA, proteins, lipids and metabolites in extracellular vesicles, Nucleic Acids Res. 47 (2018) D516–D519, https://doi.org/10.1093/nar/gky1029.
[36] S. Keerthikumar, D. Chisanga, D. Ariyaratne, H. Al Saffar, S. Anand, K. Zhao, M. Samuel, M. Pathan, M. Jois, N. Chilamkurti, L. Gangoda, S. Mathivanan, ExoCarta: a web-based compendium of exosomal cargo, J. Mol. Biol. 428 (2016) 688–692, https://doi.org/10.1016/J.JMB.2015.09.019.
[37] D.K. Kim, B. Kang, O.Y. Kim, D.S. Choi, J. Lee, S.R. Kim, G. Go, Y.J. Yoon, J.H. Kim, S.C. Jang, K.S. Park, E.J. Choi, K.P. Kim, D.M. Desiderio, Y.K. Kim, J. Lötvall, D. Hwang, Y.S. Gho, EVpedia: an integrated database of high-throughput data for systemic analyses of extracellular vesicles, J. Extracell. Vesicles 2 (2013), https://doi.org/10.3402/jev.v2i0.20384.
[38] H. Kalra, G.P.C. Drummen, S. Mathivanan, Focus on extracellular vesicles: introducing the next small big thing, Int. J. Mol. Sci. 17 (2016), https://doi.org/10.3390/ijms17020170.
[39] M. Colombo, G. Raposo, C. Théry, Biogenesis, secretion, and intercellular interactions of exosomes and other extracellular vesicles, Annu. Rev. Cell Dev. Biol. 30 (2014) 255–289, https://doi.org/10.1146/annurev-cellbio-101512-122326.
[40] J.M. Escola, M.J. Kleijmeer, W. Stoorvogel, J.M. Griffith, O. Yoshie, H.J. Geuze, Selective enrichment of tetraspan proteins on the internal vesicles of multivesicular endosomes and on exosomes secreted by human B-lymphocytes, J. Biol. Chem. 273 (1998) 20121–20127, https://doi.org/10.1074/jbc.273.32.20121.
[41] D. Perez-Hernandez, C. Gutierrez-Vazquez, I. Jorge, S. Lopez-Martin, A. Ursa, F. Sanchez-Madrid, J. Vazquez, M. Yanez-Mo, The intracellular interactome of tetraspanin-enriched microdomains reveals their function as sorting machineries to exosomes, J. Biol. Chem. 288 (17) (2013) 11649–11661, https://doi.org/10.1074/JBC.M112.445304, jbc.M112.445304.
[42] Z. Andreu, M. Yáñez-Mó, Tetraspanins in extracellular vesicle formation and function, Front. Immunol. 5 (2014) 1–12, https://doi.org/10.3389/fimmu.2014.00442.

[43] D.P. Romancino, V. Buffa, S. Caruso, I. Ferrara, S. Raccosta, A. Notaro, Y. Campos, R. Noto, V. Martorana, A. Cupane, A. Giallongo, A. d'Azzo, M. Manno, A. Bongiovanni, Palmitoylation is a post-translational modification of Alix regulating the membrane organization of exosome-like small extracellular vesicles, Biochim. Biophys. Acta Gen. Subj. 1862 (2018) 2879–2887, https://doi.org/10.1016/j.bbagen.2018.09.004.
[44] F. Vences-Catalán, S. Levy, Immune targeting of tetraspanins involved in cell invasion and metastasis, Front. Immunol. 9 (2018) 1277. https://www.frontiersin.org/article/10.3389/fimmu.2018.01277.
[45] M.S. Pols, J. Klumperman, Trafficking and function of the tetraspanin CD63, Exp. Cell Res. 315 (2009) 1584–1592, https://doi.org/10.1016/j.yexcr.2008.09.020.
[46] T. Shimada, S. Yasuda, H. Sugiura, K. Yamagata, Syntenin: PDZ protein regulating signaling pathways and cellular functions, Int. J. Mol. Sci. 20 (2019), https://doi.org/10.3390/ijms20174171.
[47] M.P. Zaborowski, K. Lee, Y.J. Na, A. Sammarco, X. Zhang, M. Iwanicki, P.S. Cheah, H.-Y. Lin, M. Zinter, C.-Y. Chou, G. Fulci, B.A. Tannous, C.P.-K. Lai, M.J. Birrer, R. Weissleder, H. Lee, X.O. Breakefield, Methods for systematic identification of membrane proteins for specific capture of cancer-derived extracellular vesicles, Cell Rep. 27 (2019) 255–268.e6, https://doi.org/10.1016/j.celrep.2019.03.003.
[48] Y. Tian, L. Ma, M. Gong, G. Su, S. Zhu, W. Zhang, S. Wang, Z. Li, C. Chen, L. Li, L. Wu, X. Yan, Protein profiling and sizing of extracellular vesicles from colorectal cancer patients via flow cytometry, ACS Nano 12 (2018) 671–680, https://doi.org/10.1021/acsnano.7b07782.
[49] S.A. Melo, L.B. Luecke, C. Kahlert, A.F. Fernandez, S.T. Gammon, J. Kaye, V.S. LeBleu, E.A. Mittendorf, J. Weitz, N. Rahbari, C. Reissfelder, C. Pilarsky, M.F. Fraga, D. Piwnica-Worms, R. Kalluri, Glypican-1 identifies cancer exosomes and detects early pancreatic cancer, Nature 523 (2015) 177–182, https://doi.org/10.1038/nature14581.
[50] C. Zurzolo, K. Simons, Glycosylphosphatidylinositol-anchored proteins: membrane organization and transport, Biochim. Biophys. Acta Biomembr. 1858 (2016) 632–639, https://doi.org/10.1016/j.bbamem.2015.12.018.
[51] S.A.A. Kooijmans, C.G. Aleza, S.R. Roffler, W.W. van Solinge, P. Vader, R.M. Schiffelers, Display of GPI-anchored anti-EGFR nanobodies on extracellular vesicles promotes tumour cell targeting, J. Extracell. Vesicles 5 (2016) 31053, https://doi.org/10.3402/jev.v5.31053.
[52] R.D. Sanderson, S.K. Bandari, I. Vlodavsky, Proteases and glycosidases on the surface of exosomes: newly discovered mechanisms for extracellular remodeling, Matrix Biol. 75–76 (2019) 160–169, https://doi.org/10.1016/j.matbio.2017.10.007.
[53] E.I. Buzás, E.Á. Tóth, B.W. Sódar, K.É. Szabó-Taylor, Molecular interactions at the surface of extracellular vesicles, Semin. Immunopathol. 40 (2018) 453–464, https://doi.org/10.1007/s00281-018-0682-0.
[54] D. van de Vlekkert, J. Demmers, X.-X. Nguyen, Y. Campos, E. Machado, I. Annunziata, H. Hu, E. Gomero, X. Qiu, A. Bongiovanni, C.A. Feghali-Bostwick, A. d'Azzo, Excessive exosome release is the pathogenic pathway linking a lysosomal deficiency to generalized fibrosis, Sci. Adv. 5 (2019) eaav3270, https://doi.org/10.1126/sciadv.aav3270.
[55] M. Nawaz, N. Shah, B. Zanetti, M. Maugeri, R. Silvestre, F. Fatima, L. Neder, H. Valadi, Extracellular vesicles and matrix remodeling enzymes: the emerging roles in extracellular matrix remodeling, progression of diseases and tissue repair, Cells 7 (2018) 167, https://doi.org/10.3390/cells7100167.

[56] C.M. Martinez-Ballesta, P. Garcia-Ibañez, L. Yepes-Molina, J.J. Rios, M. Carvajal, The expanding role of vesicles containing aquaporins, Cells 7 (2018), https://doi.org/10.3390/cells7100179.
[57] L.A. Mulcahy, R.C. Pink, D.R.F. Carter, Routes and mechanisms of extracellular vesicle uptake, J. Extracell. Vesicles 3 (2014), https://doi.org/10.3402/jev.v3.24641.
[58] M. Mathieu, L. Martin-Jaular, G. Lavieu, C. Théry, Specificities of secretion and uptake of exosomes and other extracellular vesicles for cell-to-cell communication, Nat. Cell Biol. 21 (2019) 9–17, https://doi.org/10.1038/s41556-018-0250-9.
[59] M. Tkach, J. Kowal, A.E. Zucchetti, L. Enserink, M. Jouve, D. Lankar, M. Saitakis, L. Martin-Jaular, C. Théry, Qualitative differences in T-cell activation by dendritic cell-derived extracellular vesicle subtypes, EMBO J. 36 (2017) 3012–3028, https://doi.org/10.15252/embj.201696003.
[60] A. Hoshino, B. Costa-Silva, T.-L. Shen, G. Rodrigues, A. Hashimoto, M. Tesic Mark, H. Molina, S. Kohsaka, A. Di Giannatale, S. Ceder, S. Singh, C. Williams, N. Soplop, K. Uryu, L. Pharmer, T. King, L. Bojmar, A.E. Davies, Y. Ararso, T. Zhang, H. Zhang, J. Hernandez, J.M. Weiss, V.D. Dumont-Cole, K. Kramer, L.H. Wexler, A. Narendran, G.K. Schwartz, J.H. Healey, P. Sandstrom, K. Jørgen Labori, E.H. Kure, P.M. Grandgenett, M.A. Hollingsworth, M. de Sousa, S. Kaur, M. Jain, K. Mallya, S.K. Batra, W.R. Jarnagin, M.S. Brady, O. Fodstad, V. Muller, K. Pantel, A.J. Minn, M.J. Bissell, B.A. Garcia, Y. Kang, V.K. Rajasekhar, C.M. Ghajar, I. Matei, H. Peinado, J. Bromberg, D. Lyden, Tumour exosome integrins determine organotropic metastasis, Nature 527 (2015) 329–335, https://doi.org/10.1038/nature15756.
[61] S. Götz, J.M. García-Gómez, J. Terol, T.D. Williams, S.H. Nagaraj, M.J. Nueda, M. Robles, M. Talón, J. Dopazo, A. Conesa, High-throughput functional annotation and data mining with the Blast2GO suite, Nucleic Acids Res. 36 (2008) 3420–3435, https://doi.org/10.1093/nar/gkn176.
[62] M. Ashburner, C.A. Ball, J.A. Blake, D. Botstein, H. Butler, J.M. Cherry, A.P. Davis, K. Dolinski, S.S. Dwight, J.T. Eppig, M.A. Harris, D.P. Hill, L. Issel-Tarver, A. Kasarskis, S. Lewis, J.C. Matese, J.E. Richardson, M. Ringwald, G.M. Rubin, G. Sherlock, Gene Ontology: tool for the unification of biology, Nat. Genet. 25 (2000) 25–29, https://doi.org/10.1038/75556.
[63] A.L. Mitchell, T.K. Attwood, P.C. Babbitt, M. Blum, P. Bork, A. Bridge, S.D. Brown, H.-Y. Chang, S. El-Gebali, M.I. Fraser, J. Gough, D.R. Haft, H. Huang, I. Letunic, R. Lopez, A. Luciani, F. Madeira, A. Marchler-Bauer, H. Mi, D.A. Natale, M. Necci, G. Nuka, C. Orengo, A.P. Pandurangan, T. Paysan-Lafosse, S. Pesseat, S.C. Potter, M.A. Qureshi, N.D. Rawlings, N. Redaschi, L.J. Richardson, C. Rivoire, G.A. Salazar, A. Sangrador-Vegas, C.J.A. Sigrist, I. Sillitoe, G.G. Sutton, N. Thanki, P.D. Thomas, S.C.E. Tosatto, S.-Y. Yong, R.D. Finn, InterPro in 2019: improving coverage, classification and access to protein sequence annotations, Nucleic Acids Res. 47 (2018) D351–D360, https://doi.org/10.1093/nar/gky1100.
[64] M. Pathan, S. Keerthikumar, C.-S. Ang, L. Gangoda, C.Y.J. Quek, N.A. Williamson, D. Mouradov, O.M. Sieber, R.J. Simpson, A. Salim, A. Bacic, A.F. Hill, D.A. Stroud, M.T. Ryan, J.I. Agbinya, J.M. Mariadason, A.W. Burgess, S. Mathivanan, FunRich: an open access standalone functional enrichment and interaction network analysis tool, Proteomics 15 (2015) 2597–2601, https://doi.org/10.1002/pmic.201400515.
[65] E.C. Dimmer, R.P. Huntley, D.G. Barrell, D. Binns, S. Draghici, E.B. Camon, M. Hubank, P.J. Talmud, R. Apweiler, R.C. Lovering, The gene ontology - providing a functional role in proteomic studies, Proteomics 8 (2008), https://doi.org/10.1002/pmic.200800002 n/a.

[66] Y. Liao, J. Wang, E.J. Jaehnig, Z. Shi, B. Zhang, WebGestalt 2019: gene set analysis toolkit with revamped UIs and APIs, Nucleic Acids Res. 47 (2019) W199–W205, https://doi.org/10.1093/nar/gkz401.

[67] S. Hildonen, E. Skarpen, T.G. Halvorsen, L. Reubsaet, Isolation and mass spectrometry analysis of urinary extraexosomal proteins, Sci. Rep. 6 (2016) 1–14, https://doi.org/10.1038/srep36331.

[68] L. Montermini, B. Meehan, D. Garnier, W.J. Lee, T.H. Lee, A. Guha, K. Al-Nedawi, J. Rak, Inhibition of oncogenic epidermal growth factor receptor kinase triggers release of exosome-like extracellular vesicles and impacts their phosphoprotein and DNA content, J. Biol. Chem. 290 (2015) 24534–24546, https://doi.org/10.1074/jbc.M115.679217.

[69] Y. Xiao, L. Zheng, X. Zou, J. Wang, J. Zhong, T. Zhong, Extracellular vesicles in type 2 diabetes mellitus: key roles in pathogenesis, complications, and therapy, J. Extracell. Vesicles 8 (2019) 1625677, https://doi.org/10.1080/20013078.2019.1625677.

[70] L. Statello, M. Maugeri, E. Garre, M. Nawaz, J. Wahlgren, A. Papadimitriou, C. Lundqvist, L. Lindfors, A. Collén, P. Sunnerhagen, M. Ragusa, M. Purrello, C. Di Pietro, N. Tigue, H. Valadi, Identification of RNA-binding proteins in exosomes capable of interacting with different types of RNA: RBP-facilitated transport of RNAs into exosomes, PloS One 13 (2018), https://doi.org/10.1371/journal.pone.0195969.

[71] H. Sork, G. Corso, K. Krjutskov, H.J. Johansson, J.Z. Nordin, O.P.B. Wiklander, Y.X.F. Lee, J.O. Westholm, J. Lehtiö, M.J.A. Wood, I. Mäger, S. EL Andaloussi, Heterogeneity and interplay of the extracellular vesicle small RNA transcriptome and proteome, Sci. Rep. 8 (2018) 10813, https://doi.org/10.1038/s41598-018-28485-9.

[72] T.A.P. Driedonks, E.N.M. Nolte-'t Hoen, Circulating Y-RNAs in extracellular vesicles and ribonucleoprotein complexes; implications for the immune system, Front. Immunol. 9 (2019) 3164. https://www.frontiersin.org/article/10.3389/fimmu.2018.03164.

[73] M.J. Shurtleff, M.M. Temoche-Diaz, K. V Karfilis, S. Ri, R. Schekman, Y-box protein 1 is required to sort microRNAs into exosomes in cells and in a cell-free reaction, Elife 5 (2016), https://doi.org/10.7554/eLife.19276.

[74] M.R. Medapati, A. Singh, R.R. Korupally, D. Henderson, T. Klonisch, S.V. Manda, P. Chelikani, Characterization of GPCRs in extracellular vesicle (EV), in: Methods Cell Biol., Academic Press, 2017, pp. 119–132, https://doi.org/10.1016/bs.mcb.2017.07.004.

[75] H. Pollet, L. Conrard, A.-S. Cloos, D. Tyteca, Plasma membrane lipid domains as platforms for vesicle biogenesis and shedding? Biomolecules 8 (2018) https://doi.org/10.3390/biom8030094.

[76] M. Record, S. Silvente-Poirot, M. Poirot, M.J.O. Wakelam, Extracellular vesicles: lipids as key components of their biogenesis and functions, J. Lipid Res. 59 (2018) 1316–1324, https://doi.org/10.1194/jlr.E086173.

[77] D.K. Kim, J. Lee, S.R. Kim, D.S. Choi, et al., EVpedia: a community web portal for extracellular vesicles research, Bioinformatics 31 (2015) 933–939, https://doi.org/10.1093/bioinformatics/btu741.

[78] P. Kroll, K. Eilers, J. Fricke, C. Herwig, Impact of cell lysis on the description of cell growth and death in cell culture, Eng. Life Sci. 17 (2017) 440–447, https://doi.org/10.1002/elsc.201600088.

[79] D. Božič, S. Sitar, I. Junkar, R. Štukelj, M. Pajnič, E. Žagar, V. Kralj-Iglič, K. Kogej, Viscosity of plasma as a key factor in assessment of extracellular vesicles by light scattering, Cells 8 (2019), https://doi.org/10.3390/cells8091046.

[80] M.C. Pahl, O.L. Askinazi, C. Hamilton, I. Cheng, K. Cichewicz, J. Kuhn, S. Manohar, C.D. Deppmann, Signalling via Single-Pass Transmembrane Proteins, ELS, 2013, https://doi.org/10.1002/9780470015902.a0025160.
[81] A.E. Russell, A. Sneider, K.W. Witwer, P. Bergese, S.N. Bhattacharyya, A. Cocks, E. Cocucci, U. Erdbrügger, J.M. Falcon-Perez, D.W. Freeman, T.M. Gallagher, S. Hu, Y. Huang, S.M. Jay, S. ichi Kano, G. Lavieu, A. Leszczynska, A.M. Llorente, Q. Lu, V. Mahairaki, D.C. Muth, N.N. Hooten, M. Ostrowski, I. Prada, S. Sahoo, T.H. Schøyen, L. Sheng, D. Tesch, G. Van Niel, R.E. Vandenbroucke, F.J. Verweij, A.V. Villar, M. Wauben, A.M. Wehman, H. Yin, D.R.F. Carter, P. Vader, Biological membranes in EV biogenesis, stability, uptake, and cargo transfer: an ISEV position paper arising from the ISEV membranes and EVs workshop, J. Extracell. Vesicles 8 (2019), https://doi.org/10.1080/20013078.2019.1684862.
[82] L. Cheng, W. Zhao, A.F. Hill, Exosomes and their role in the intercellular trafficking of normal and disease associated prion proteins, Mol. Aspect. Med. 60 (2018) 62–68, https://doi.org/10.1016/j.mam.2017.11.011.
[83] L.S. Watson, E.D. Hamlett, T.D. Stone, C. Sims-Robinson, Neuronally derived extracellular vesicles: an emerging tool for understanding Alzheimer's disease, Mol. Neurodegener. 14 (2019) 22, https://doi.org/10.1186/s13024-019-0317-5.
[84] P.D. Robbins, A.E. Morelli, Regulation of immune responses by extracellular vesicles, Nat. Rev. Immunol. 14 (2014) 195–208, https://doi.org/10.1038/nri3622.
[85] C. Théry, M. Ostrowski, E. Segura, Membrane vesicles as conveyors of immune responses, Nat. Rev. Immunol. 9 (2009) 581–593, https://doi.org/10.1038/nri2567.
[86] B.D. Angst, C. Marcozzi, A.I. Magee, The cadherin superfamily: diversity in form and function, J. Cell Sci. 114 (2001) 629 LP–641. http://jcs.biologists.org/content/114/4/629.abstract.
[87] B. Wang, Z. Tan, F. Guan, Tumor-derived exosomes mediate the instability of cadherins and promote tumor progression, Int. J. Mol. Sci. 20 (2019) 3652, https://doi.org/10.3390/ijms20153652.

CHAPTER FOUR

Post-translational lipidation in extracellular vesicles: chemical mechanisms, biological functions and applications

Sabrina Picciotto[a], Daniele P. Romancino[a], Valentina Buffa[a,b], Antonella Cusimano[a], Antonella Bongiovanni[a,]* and Giorgia Adamo[a,]*

[a]Institute for Biomedical Research and Innovation (IRIB), National Research Council of Italy (CNR), Palermo, Italy

[b]Integrare UMR_S951Genethon, Inserm, University of Evry, Université Paris Saclay Genethon, Evry, France

*Corresponding author: E-mails: antonella.bongiovanni@irib.cnr.it, giorgia.adamo@irib.cnr.it

Contents

Advances in Biomembranes and Lipid Self-Assembly, Volume 32
ISSN 2451-9634
https://doi.org/10.1016/bs.abl.2020.05.001

Abstract

Extracellular vesicles (EVs) are phospholipid bilayer enclosed structures containing specific proteins, nucleic acids, and metabolites, which are naturally released from cells as vehicles of intercellular communication. Post-translational modifications (PTMs) of EV proteins can regulate EVs biogenesis or play diverse roles in their biology, in both homeostatic and pathological conditions. Here, we focus on the mechanism and biological roles of post-translational lipidations of EV proteins, with a particular insight on their involvement in EV biogenesis and functions. We investigate various aspects of PTM by lipidation in depth, ranging from chemistry to technical advances for lipidation detection and recent biotech applications. Among lipidations on EV proteins, palmitoylation is the most widespread mechanism. EVs have been shown to be enriched of palmitoylated proteins, and the profile of EV palmitoyl proteome could contribute to early diagnosis in some kinds of cancer, suggesting that EV protein lipidation can be exploited as potential early tumor biomarker. Thus, the emerging evidence on EV protein lipidations could help us to better understand their physio/pathological function with implications for preclinical and clinical settings.

1. Introduction

The production of different types of membrane-confined vesicles, called extracellular vesicles (EVs), is a very well conserved cellular process occurring in all three domains of life, from Archaea to Bacteria and Eukarya [1]. EVs are delimited by a lipid bilayer and contain proteins, nucleic acids, and lipids that are exchanged by the cells establishing a cell-cell communication system to maintain healthy cellular homeostasis or to signal a pathological state [2]. EVs include heterogeneous, yet overlapping, subpopulations in terms of size, density and even molecular composition; they have been broadly divided into two categories: small extracellular vesicles (sEVs) ranging from 50 to 200 nm in diameter and including exosomes, and large extracellular vesicles (lEVs) ranging from 100 to 10,000 nm. Several proteins have been indicated as exosome markers, such as Alix (also referred to as programmed cell death 6-interacting protein, PDCD6IP), the tetraspanin transmembrane proteins CD9, CD63, and CD81, heat-shock 70 kDa proteins (HSP70, HSC70), flotilin 1, and major histocompatibility complex

(MHC) class I and class II proteins [3,4]. EVs can originate either from the plasma membrane surface through budding and the subsequent release of the vesicle into the extracellular space, or through inward budding of the endosomal membrane to form intraluminal vesicles (ILVs) inside multivesicular endosomes (MVEs), with subsequent MVE fusion with the membrane surface and release of the ILVs outside the cell. Although the physical similarities between EVs originating from the two biogenic pathways has hindered their characterization, recent data indicates that sEVs (e.g., exosomes) encapsulate distinct types of proteins and at different expression levels, and that they likely enact different biological functions [5]. In line with these findings, accumulating evidence supports the idea that EV biogenesis and cargo selection is a finely-tuned process and that the molecular composition and rates of production of sEVs and lEVs vary greatly, depending on cell-type and physiological state [6,7].

Post-translational modifications (PTMs) can affect a protein's structure and function by changing its interactors and enzymatic activity. At present, more than 450 PTMs have been annotated in the Uniprot database [8]. Although the pathways that control EV cargo selection and packaging remain poorly understood, recent findings have begun to shed light on some of the molecular mechanisms that specify the proteins to be secreted into EVs. In this context, there has been an emerging interest in the role of PTMs in EV biology and several proteomic and cell biological studies have investigated the role of PTMs in EV biology. Proteins which have been routed into sEVs also undergo PTM through processes such as phosphorylation, acetylation, lipidation, and ubiquitination, and there are a number of excellent reviews covering PTMs including phosphorylation, acetylation, oxidation, ubiquitination, and citrullination [9,10]. In this chapter, we focus on PTMs of protein by lipidation and introduce recent findings regarding the lipid-related protein sorting pathway into sEVs. Fatty acid protein modification includes different types of lipidation, such as myristoylation, prenylation, and palmitoylation. These lipid modifications play key roles in different functions associated with subcellular localization, membrane-protein and protein-protein interaction, and protein structure and stability [11]. There is a correlation between protein lipidation and EVs, originating from the membrane surface or endosomal system of the cells [12]. However, the molecular mechanisms driving the lipidated proteins into EVs remain to be fully elucidated. Certainly, lipidation and regulation of EV proteins constitute a code of cellular communication to respond in a fast, versatile, and precise way [13]. The following sections discuss the biological function and application of protein lipidation in EVs.

2. Exploring protein lipidation: chemical modifications and functions

2.1 Glycosylphosphatidylinositol anchor (GPI)

This PTM consists of the covalent attachment of a glycosylphosphatidylinositol (GPI) to the C-terminal of proteins. It takes place in the ER, and the modified protein is subsequently translocated and anchored to the outer leaflet of the plasma membrane [14]. The common structure of GPI anchors, conserved in eukaryotes, is made of an ethanolamine phosphate (EtNP), three mannoses (Mans), one non-N-acetylated glucosamine (GlcN), and an inositol phospholipid. However, it can be modified differently in diverse organisms or cell types. For example, side-chain modifications can be attached at the mannoses, as well as various lipid chains at the phosphoinositol (diacylglycerol, alkylacylglycerol, arachidonic acid, etc.), resulting in increased complexity [15—17]. Besides targeting proteins to the membrane, the GPI PTM has many other biological roles, especially concerning cell-cell contact and adhesion processes. It has been shown to be involved in physio/pathological processes such as immune response, inflammation, and oncogenesis [18,19].

2.2 Prenylation (farnesylation and geranylation)

Prenylation is an irreversible protein PTM consisting of the addition of multiple isoprene units to the cysteine located on the C-terminal of proteins. It is an enzymatic process for which the consensus motif is known as the CAAX box (C for cysteine, A for an aliphatic amino acid, and X for any amino acid). If the X is a leucine, the Geranygeranyl Transferase type I will add a 20-carbon isoprenoid geranylgeranyl; otherwise, the Farnesyl Transferase will attach a 15-carbon farnesyl to the cysteine [20]. Once proteins are prenylated in the cytoplasm, the AAX tripeptide is cleaved in the ER by a protease (Ras-converting enzyme 1 or a-factor converting enzyme 1) and, subsequently, the C-terminal isoprenylated cysteine is methylated by an isoprenylcysteine carboxylmethyltransferase (ICMT) [21]. These two final processes promote protein-membrane association, even though stable membrane binding is usually mediated by the combination with other membrane targeting modifications [22]. Typical proteins showing the

CAAX sequence are members of the G-protein superfamily, and most of them, such as Ras, Rho, Rac, and CDc42, are associated with the membranes [23]. Rab proteins, instead, contain the motif CC or CXC at the C-terminal, and thus are prenylated by the Rab Geranylgeranyl Transferase (RabGGTase), a type 2 GGTase [24]. It is known that several cellular processes depend on proper protein targeting to cellular membranes; accordingly, Ras GTPase targeting can influence cellular growth and migration. Therefore, in a cancer cell context, the prenylation of oncogenic Ras appears to be determinant for its activity [25].

2.3 N-terminal myristoylation

N-myristoylation consists of the covalent attachment of a saturated 14-carbon fatty acyl group, myristoyl, to the N-terminal glycine of a protein via an amide bond, and it is catalyzed by an N-myristoyl transferase (NMT) [26]. This reaction occurs predominately co-translationally, by the recognition of the consensus motif Gly-XXX-Ser/Thr/Cys after the cleavage of the initiating methionine by a methionine aminopeptidase [27]. However, this modification can also occur post-translationally; this is the case during apoptosis when a caspase cleavage promotes the exposition of an internal glycine. In the apoptotic context for proteins such as Bid and p21-activated protein kinase 2 (PAK2), this modification is crucial for correct protein functioning [28–30].

However, the addition of myrystate is not sufficient to anchor proteins to lipid membranes. Accordingly, myristoylation often acts together with other mechanisms to guide protein targeting and functions. These mechanisms include electrostatic interactions, ligand binding, or the occurrence in the same protein of another modification, such as palmitoylation [31–33].

N-myristoylation concerns different kind of proteins, cellular proteins members of the signal transduction pathway, such as Src, as well as proteins of viral origin, including Gag or Nef. In the latter case, the virus takes advantage of the cellular NMT to improve protein functions and increase its virulence [34,35].

2.4 C-terminal cholesterol esterification

Lipidation involving the attachment of a cholesterol molecule has been identified so far in the protein members of the Hedgehog family. It is the result of a peculiar autocleavage process that leads to the incorporation of cholesterol with the release of the C-terminal protein domain. This process is similar to that of intein self-splicing proteins [36] and consists of two steps.

First, the sulfhydryl group of the N-terminal cysteine attacks the carbonyl of the C-terminal glycine residue to form a thioester intermediate. Second, the cholesterol 3ß-hydroxyl group attacks this thioester linkage [37] forming an oxyester bond and releasing the C-terminal domain, that will undergo for ER-associated degradation [38].

Hedgehog proteins are not only cholesterylated, but also palmitoylated at the N-terminal cysteine, and both lipidations are critical for the protein's functioning. Removal of cholesterol leads to an inactive form of the protein [38] and, consequently, to an impairment of Hedgehog ligands distribution. The presence of cholesterol [39] on the other hand allows the formation of multimeric complexes [40] and mediates the interaction with lipoprotein particles [41].

2.5 Lysine fatty acylation (myristoylation and palmitoylation)

Lysine residues of proteins can be a site for acylation, especially with the use of metabolically-derived components, such as the different kinds of acyl-CoA, as the source of acyl groups. The first mammalian protein discovered to have an acylated lysine was TNFalpha (myristoylated on Lys19 and Lys20) [42]. Lysine myristoylation has also been shown as well in interleukin1a [43], while in acquaporin-0 lysine palmitoylation has been observed [44]. Lysine acylation can regulate the availability and degradation of the protein. The presence of lysine myristoylation in TFNalpha promotes its targeting and degradation into lysosomes; therefore, the action of defatty-acylation (by Sirtuin proteins) allows TFNalpha secretion [45]. Lysine acylation has other key biological roles; for instance, this lipidation facilitates plasma membrane targeting of the protein R-Ras2, in this way tuning its signaling pathway and the consequent regulation of cell proliferation [46].

2.6 Serin fatty acylation (serin fatty octanoylation)

Octanoylation consists of the attachment of an octane, an eight-carbon fatty acid, to a serine residue through an oxyester linkage. The only peptide known to be modified with this lipidation in mammals is ghrelin. This 28-residue peptide, once acylated, is able to bind to the growth hormone secretagogue receptor (GHSR) regulating growth hormone release and energy homeostasis, while de-acyl ghrelin antagonizes this effect [47]. The reaction of octanoylation is catalyzed by ghrelin octanoyltransferase (GOAT), which adds the fatty acid to serine 3 within the N-terminal sequence substrate recognition [48].

2.7 Cystein and serin palmitoylation

The most common acylation involving cysteine residues is the attachment of the 16-carbon saturated palmitic acid via a reversible thioester linkage also known as S-palmitoylation [49]. However, the N-palmitoylation (characterized by a stable amide bond) of N-terminal cysteines of proteins, such as the alpha-subunit of the heterotrimeric G protein [50] and Sonic Hedgehog (Shh), has also been reported [51]. The unique feature of S-palmitoylation is its reversibility that allows the dynamic and fine regulation of the target protein's hydrophobicity. This promotes a strong association with the lipid bilayer, especially in proteins lacking a transmembrane domain. Besides its key role in influencing membrane affinity and targeting, palmitoylation has also been shown to be critical for subcellular protein trafficking, protein stability, modulation of protein interactions (i.e., by altering the accessibility of this region to other PTMs), particularly affecting adhesion, signaling, and organelle structure [52—54]. S-palmitoylation of proteins is mediated by palmitoyl transferases (PATs), multi-pass membrane proteins also called DHHC palmitoyl acyltransferases, for the highly conserved Asp-His-His-Cys sequence in the catalytic site [55]. The reaction that results in S-palmitoylation does not require any energy source since the palmitoyl donor, palmitoyl-CoA, contains a thioester bond equivalent in terms of energy to the quantity necessary for cysteine palmitoylation. The enzyme uses a "ping pong mechanism", made of two steps: in the first, the DHHC is autoacylated, forming an acyl-enzyme intermediate that, in the second step, transfers the palmitoyl to a cysteine residue of the targeted protein substrate [56]. The reversibility of the process is ensured by palmitoyl thioesterases, enzymes able to cleave the palmitate from S-palmitoylated proteins [57].

Serine O-acylation, another reported protein lipid acylation, plays a key role in Wnt functions. Wnt proteins are secreted glycoproteins which are important in several aspects of embryonic development as well as adult tissue self-renewal. Wnt acylation consists of the attachment of a monounsaturated form of palmitate, cis-Δ9-palmitoleate (cis-Δ9-16:1), via an oxyester bond to a conserved serine residue. Wnt protein palmitoylation is mediated by the Porcupine (Porcn) enzyme, a member of the family of transmembrane-bound O— acyltransferases (MBOAT), after recognition of the Wnt consensus sequence surrounding the conserved cysteine [58]. This modification is required for Wnt proteins to bind to the Wnt receptor Frizzled [59]. Moreover, it is crucial as well for Wnt secretion since it allows interaction with Wntless, an intracellular transmembrane protein that facilitates Wnt migration from the Golgi apparatus to the plasma membrane [60].

It has also recently been reported that also Wnt acylation, such as S-palmitoylation, is reversible, thanks to the specific cleavage activity of the hydrolase Notum. This enables the inactivation of Wnt and thus provides for the fine regulation of its signaling [61].

3. Methodological and technical advances for detecting lipid PTMs

Several biochemical and molecular biological methods are available for investigating lipid-based PTMs of proteins. The majority of lipidation detection protocols have been developed in combination with enrichment strategies because of the low abundance of lipid modified proteins, thus requiring an enrichment step.

3.1 Radioactive isotope-labeling of precursor lipids

The classical and common methods used for detecting lipidated proteins is based on the incorporation of radioactive precursors and the subsequent detection of the modified proteins. For example, in the detection of prenylated proteins, a radioactive precursor such as [3H]-mevalonic acid is used, which is incorporated into the cell, taking active part in the biosynthesis of cholesterol [62]; otherwise, ^{3}H-myristic acid is used.

Examples of these metabolic labeling methods are the study performed on gelsolin [63] and actin PTMs [89]. This method is time-consuming, needs expensive precursors, and is hazardous to human health; however, many approaches have been developed to optimize the protocol.

3.2 Mass spectrometry

Mass spectrometry (MS)-based methodologies allow for large-scale protein PTM analyses, including lipidation, In general, the proteins extracted from the cell or tissue are separated by SDS page gel electrophoresis or two-dimensional gel electrophoresis (2DE); then a mix of peptides is produced by enzyme digestion that is separated by liquid chromatography and directly ionized. The ionization of the peptide is done with either electrospray ionization (ESI) or matrix-assisted laser desorption ionization (MALDI). Afterward, the ionized molecules formed are separated based on their mass to charge ratio, toward mass analyzers, like time of flight (TOF) or linear ion trap (IT). The separated ions are then detected and form peaks in the mass spectrum. Further fragmentation of the ions of interest is done using various methods, such as collision induced dissociation (CID) or electron transfer dissociation (ETD).

However, due to the hydrophobicity of the lipidated samples, the recovery and detection of peptides is often low. To overcome this problem, new sample preparation methods have been developed for both myristoylated [64,65] and palmitoylated peptide or protein detection [66]. Furthermore, lipidated samples can be analyzed by HPLC-MS using the Multidimensional protein identification technology (MudPIT) [67]. This approach is a non-gel technique based on 2-dimensional chromatography separation of the peptides (a cation exchange column followed by a reverse phase chromatography column) prior to ionization. One of the advantages of this technique is that it is compatible with hydrophobic or membrane proteins such as acylated proteins. A new version of these methods, modified by Casey and co-workers [68], uses a resin-assisted capture (RAC) [69] coupled with MS-based proteomics to detect S-nitrosylation sites in proteins.

An alternative method is based on acid esterification after the cleavage of the thioester group, releasing the fatty acid and hydrogenation in the presence of formic acid and ethanol [70]. However, this approach cannot be used for proteins with low expression levels or EV proteins since it needs a high quantity (1 μg) of purified proteins. Moreover, it only gives information about the acyl groups present in the protein, but not about the acylation site.

3.3 Acyl-biotin exchange

Acyl-biotin exchange (ABE), developed by Drisdel and Green in 2004, is a non-radioactive method for the detection of protein palmitoylation [71]. The ABE assay consists of three main steps: (1) treatment of the sample with N-ethylmaleimide (NEM), which results in the irreversible blockade of unmodified cysteine thiol groups; (2) specific cleavage of palmitate from cysteine by hydroxylamine (HAM), a powerful reducing agent, which results in the unmasking of cysteines which are palmitoylation sites; (3) selective labeling of the palmitoylated cysteine using biotin-BMCC, a maleimide conjugated biotin molecule that has a high affinity for free thiol groups of cysteines. Finally, purified and eluted thiol-biotinylated proteins are analyzed by SDS-PAGE and detected by streptavidin-HRP antibody in a western blotting assay [72].

Based on ABE method, a slightly different procedure has been developed to detect protein palmitoylation in sEVs [73]. Palmitoylation is detected directly on the immunoprecipitated protein of interest; the three steps, described above, are indeed performed directly on the PVDF-immobilized protein and not in solution [73].

3.4 Bioorthogonal probes

Bioorthogonal probes are azide- or alkyne-functionalized palmitate analogs, e.g., 14-azidotetradecanoic acid (azido-palmitate) [74] that, when added to cell culture medium, are incorporated into native S-acylation sites by the endogenous S-acylation enzymatic system. Newly labeled proteins are thus synthesized and, after lysis, can be conjugated to a biotin analog (e.g., phosphin-biotin [98], biotin-azide [75], or azido-azo-biotin [76]) by click chemistry. Then biotinylated S-acylated proteins can be purified by streptavidin affinity.

This approach is essentially a metabolic labeling with lipid analogs containing chemical groups which are not present in biological samples (mostly azide and alkyne modifications) that become substrates for bioorthogonal reactions with chemical tags that allow them to be studied in their native settings. Two different bioorthogonal reactions have been mainly employed, the Staudinger ligation and the copper-catalyzed azide—alkyne 1,3-dipolar cycloaddition (CuAAC) of azides with terminal alkynes (Huisgen reaction or click chemistry) [77].

The Staudinger reaction is a chemical reaction of an azide with a phosphine or phosphite, producing an iminophosphorane, and it is named the Staudinger ligation because it ligates two molecules together. In studies of protein lipid modification, the researcher successfully used azido-containing farnesyl substrates for protein labeling in cells [78]. Based on the reaction of an azide with a modified triaryl phosphine to form a stable amide bond [79], the resulting prenylated proteins are incubated with a biotinylated phosphine-capture reagent and collected by streptavidin beads [80,81].

CuAAC: is a method of protein palmitoylation profiling using an alkyne derived palmitic acid [82]. CuACC is also based on the azide reaction but, in this case, with a terminal alkyne. Through metabolic labeling with this alkyne-fatty acid reaction via CuAAC with a biotin-azide, proteins were enriched and subjected to proteomic analysis using MudPIT technology; this uses copper as a catalyst, which can be toxic for living cells [83] and result in protein denaturation [84]. Nevertheless, a copper-free approach has been developed involving the use of cyclooctyne, which may overcome this limitation [85].

Finally, a quick reference to classical antibody-based methods such as western blotting, ELISA, and microscopy techniques. Most rely on the availability of antibodies that can specifically recognize a modified amino acid residue within a protein. These techniques can provide information about cellular functions, spatial subcellular distribution, and enzymatic activity, all of the characteristics of proteins that can be modified by PTMs.

In order to be detected and considering their low abundance, lipid-modified proteins usually require an enrichment step. In combination with enrichment strategies, specific modified protocols are required to perform lipidation detections in EV membranes, adapted to the EV characteristics (e.g. low lipidated-protein abundance and EV yields, and small EV size). In this regard, a modified ABE assay, performed on immunoprecipitated proteins immobilized on PVDF membranes, was developed by Romancino et al. in order to investigate S-palmitoylation of Alix in EVs [73]. In the same time, Mariscal et al. combined low-background acyl-biotinyl exchange (LB-ABE) with label-free proteomics in order to analyze the palmitoyl proteome of EVs prostate cancer cells [86]. The strategy combining the metabolic labeling of newly synthesized EV proteins with active azides and bioorthogonal click conjugation performed for EV targeting, has been successful as well [87]. Lastly, referring to the small size of EVs only Electron Microscopy, both TEM (transmission electron microscopy) and SEM (scanning electron microscopy) [88,89] are able to produce high-resolution images of submicron particles, such as exosomes, as much as super-resolution microscopy allows immunofluorescence analysis [90].

3.5 *In silico* tools for the prediction of protein lipidation

Different computational approaches exist to perform preliminary analysis and predict PTM sites that can be further investigated with experimental techniques. The *in-silico* tools available, often complemented with molecular modeling and simulations, allow to predict protein PTMs, in term of lipidation sites, and could provide a better understanding of the biological function of PTMs. To predict N-myristoylation sites, NMT and Myristoylator can be used [91,92]. In particular, Myristoylator predicts myristoylation only on terminal glycines and is highly specific. The online tool PrePS is based on modeling of the substrate-enzyme interactions; it predicts protein CAAX farnesylation, CAAX geranylgeranylation and Rab geranylgeranylation sites [93]. For palmitoylation, several tools have been developed; the first was CSS-palm that, together with CKSAAP-PALM, is largely used for the prediction of palmitoylated sites [94,95]. SwissPalm compares different palmitoyl-proteomic studies and gives the possibility to search for a protein of interest, looking for predicted S-palmitoylation sites and identifying orthologues [96]. Besides the prediction of palmitoylated sites, the tool SeqPalm correlates the mutation of palmitoylation sites with diseases [97]. Recently, it has been developed a new predictor called GPS-Palm based on a deep learning-based graphic presentation system; starting form a large data set of 3098 unique S-palmitoylation sites, it increases the accuracy for the prediction of S-palmitoylation sites, and offers also species-specific predictors for human- and mouse-specific sites [98]. In addition to the described systems, GPS-lipid is an *in-silico* tool that allows the prediction of multiple lipid modification sites at the same time, and this can represent a great help to study co-regulation between different types of lipidations [99].

4. Lipidation of EV proteins in cellular physiology and disease

Different cytosolic proteins are linked to the inner leaflet of the plasma membrane, before being assigned to EVs. This interaction relies on protein PTMs, such as palmitoylation, prenylation and myristoylation, which anchor and concentrate cargo proteins to small membrane domains, from which, likely, EVs will bud [2].

The effects of protein lipidation on cellular functions are achieved by regulating protein–membrane interactions and, perhaps somewhat surprising, protein–protein interactions, stability, and enzymatic activities. In EVs, deciphering the role of protein lipidations (see Table 1) might help in understanding the cargo sorting mechanisms and to identify novel functions of this PTM in health and disease [12,100].

It should be noted that the majority of the protein lipidations reported in the previous paragraph (2.) were not identified in EV preparations and more studies are necessary to validate each modification in EVs Nevertheless, palmitoylation is an important regulator of sEV membrane organization and cargo localization. Indeed, by a comparative analysis of publicly available palmitoylation (SwissPalm) and sEV databases (ExoCarta), we have demonstrated that sEVs are three times enriched in S-palmitoylated proteins compared to the human proteome. More striking, we have evaluated the distribution of S-palmitoylated proteins across different cellular compartments using gene ontology (GO) annotation for cellular components, and consistent with the previous findings, exosomes contained the highest percentage of S-palmitoylated proteins (80.7%, as determined considering the

Table 1 Comparison between structure, reversibility and identification method in different protein EV-associated lipidation: palmitoylation, myristoylation, prenylation, GPI and C-terminal cholesterol esterification.

Lipidation	Structure	Reversibility	EV source	Examples
Palmitoylation	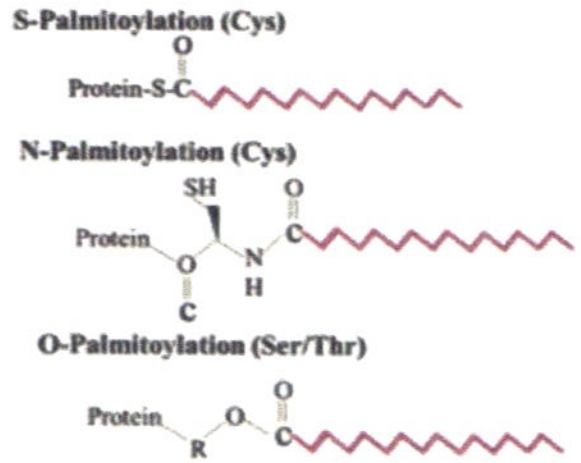	Reversible	*D. melanogaster*	Wnt [107],
			Mouse myoblast cells	Alix [73],
		Irreversible	*Human blood plasma, amniotic fluid, urine, saliva, breast milk*	TEMs [111],
		Reversible	*Human breast cancer cells*	TLR2 [121],
			Rat hippocampal neurons	Shh [108].
N-Myristoylation			*C. elegans*	CIL-7 [106],
		Irreversible	*Drosophila S2 cells*	TGs [109],
			Human Jurkat T cells	TyA [144].
Prenylation		Reversible	*Human Jurkat T cells*	TyA [144].
Glycosylphosphatidyl-inositol anchor (GPI)	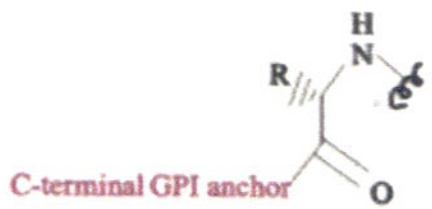		*Human adipose tissues*	T-cadherin [145],
		Reversible	*Different human cancer cell lines*	DRMs [146],
			Human melanoma cells	IL2 [147],
C-terminal cholesterol esterification	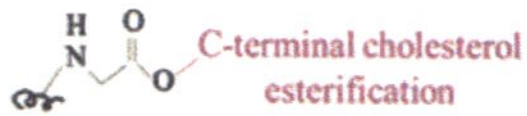	Reversible	*Rat hippocampal neurons*	Shh [108].

ExoCarta dataset), compared to all the other cellular compartments, including plasma membrane (35%) and other membrane-enclosed intracellular organelles, as lysosomes (2.1%) and endosome (3.2%). The high percentage of palmitoylated EV proteins suggests that this PTM is a distinctive signature of exosomal proteins [101,102]. Thus, paragraphs below will report recent findings on EV protein-related lipidations, focusing on palmitoylation and its role in cellular physiology and human disease.

4.1 Lipidation of EV proteins: its role in cellular physiology

Lipidations are involved in the binding of proteins to the membrane, in their endocytic trafficking, and to promote interactions with transporter proteins [103]. In this context, it has been reported that, in *Caenorhabditis elegans* and *Drosophila melanogaster,* EVs carry lipidated morphogens which are important for tissue remodeling, especially for epithelial and neural cells [104–106]. In *C. elegans*, the EVs are involved in the physiological development of the cuticle, which is composed primarily of glycosylated and lipid-modified proteins secreted by the underlying epithelial cells [105]. The loss of **CIL-7** morphogen causes defective mating behavior in males, due to the accumulation of EVs in the lumen of the cephalic sensory organ [106]. The authors also showed that myristoylation is important for the CIL-7 ciliary protein's functioning and targeting to EVs [106]. Furthermore, EVs play a key role in synaptic growth at the neuromuscular junction (NMJ) in *Drosophila* larvae [104]. Neurons at the NMJ release EVs carrying the lipidated morphogen **Wingless (Wg)**, orthologous to mouse Wnt proteins, and the Wg carrier protein Wntless (Wls). The Authors reported that upon specific palmitoylation (O-palmitoleoylation) on conserved serine, Wnt proteins become associated with Wls. Other studies supported this notion, showing that Wls is also loaded in EVs and released from cells [104,107].

Another example of lipidation linked to EV cargo sorting is given by the regulation of the **Sonic hedgehog (Shh)** signaling pathway that plays critical roles in embryonic development as well as in adult tissue homeostasis. In the rat cerebellum from postnatal day P(2), Shh produced by Purkinje cells stimulates the growth of granule neuron precursor cells. In the mature cerebellum, neuron-derived Shh continues to function by determining the molecular features of neighboring glial cells. The bioactivity of Shh depends on dual-specific lipid modifications: the addition of palmitate at its N-terminal and cholesterol at its C-terminal. These lipidations make Shh molecules lipophilic, preventing them from diffusing freely in the extracellular space. Indeed, studies have shown that the lipidate-Shh is transported at the intercellular level by EVs, allowing the protein to perform its function in the specific extracellular district [108]. These findings highlight the importance of EV-associated protein palmitoylation, in particular for the targeting of many neuronal proteins at specific region and its key role in synaptic plasticity.

A further finding derives from studies on *Drosophila* **transglutaminases (TGs)**, which are Ca^{++}- dependent enzymatic proteins that catalyze protein cross-linking by lysine and glutamine residues to form E-(γ-glutamyl) lysine bonds; these derived protein aggregates have a role in blood clotting and bone formations. In *Drosophila*, the TG-A isoform is N-terminal myristoylated and S-palmitoylated, co-localized with transfected tetraspanin CD63 and Rab7 proteins (both MVE markers) into MVEs, and is secreted by exosomes. The TG-A fatty acylation has been shown to be essential for exosome secretion, given that the inhibition of S-palmitoylation using 2-bromopalmitate (2BP) prevented TG-A positive exosomes secretion [109].

Focusing on the EV biogenesis, it is known that palmitoylation regulates the tetraspanins (including **CD9 and CD63)** function. Tetraspanins are organized in membrane microdomains, termed tetraspanin-enriched microdomains (TEMs), by forming clusters from which likely EVs originate; S-palmitoylation contributes to the supramolecular organization of tetraspanins in these membrane domains [110,111]. Furthermore, **Alix**, a protein also involved in viral/EV biogenesis, was found to be S-palmitoylated in muscle cells, and the inhibition of palmitoylation, using 2BP, negatively influenced Alix-CD9 interaction, Accordingly, inhibition of palmitoylation also affected cargo distribution of Alix, CD9, and CD63 into sEVs and increased the number, the size and the membrane structure of the sEVs [73]. In the context of muscle cells, a recent finding showed that EV secretion occurs at massive level during myofiber repair and regeneration [112]. Indeed, after myofibers are injured, a plasma membrane repair process occurs [113,114]. This event is mediated by shedding of Alix-positive EVs at the site of injury [114]. This finding highlights the key role of Alix in skeletal muscle; however, the involvement of palmitoylation in muscle repair as well as other EV-related physiological events remain to be elucidated. In this context, we have recently reported a key role of S-palmitoylation in maintaining a proper membrane structure organization of sEVs, in addition to regulating the proper localization of Alix in SkM sEVs and supporting the interactions among exosome specific regulators/biomarkers (i.e., Alix and CD9) [73].

4.2 Lipidation of EV proteins in human diseases

Lipid-based PTMs in EVs have been reported in physiology and in different pathologies, such as cancer, autoimmunity, atherosclerosis and neurodegenerative diseases and studies of EV protein lipidation are important to understand many human diseases [115]. An example of a pathology that has been linked to lipidation is the vascular calcification that leads to cardiovascular disease. Vascular calcification is mediated by high expression levels of sortilin, and the subsequent sortilin-dependent targeting of tissue-nonspecific alkaline phosphatase (TNAP) to EVs [116]. Sortilin is a protein receptor that binds and controls target proteins, such as TNAP, sorting along the secretory or endocytic compartments. In this view, sortilin protein trafficking is key in this process; it can be found into two forms: (i) homodimers linked by an intermolecular disulfide bond at Cysteine783 (Cys783) residue, which is regulated by ligand binding in the extracellular domain, and (ii) monomers with Cys783 palmitoylated, and not available for disulfide bond. Thus, homodimers of sortilin are directed to the EVs, whereas palmitate monomers are carried to the Golgi apparatus. Inhibition of palmitoylation using 2BP results in increase of homodimers, and subsequently sortilin trafficking to the EVs [117]. Finally, potentially inhibition of palmitoylation could be used to control sortilin/TNAP trafficking to EVs and EV-associated diseases, such as vascular calcification. This finding is one of the few example of the negative regulation of EV cargo sorting by S-palmitoylation.

In contrast, it has been shown the direct link between palmitoylated proteins and their accumulation on the surface of sEVs during cancer progression [118,119]. A large-scale analysis of the palmitoyl-proteome from sEVs and lEVs, derived from prostate cancer cell showed that three proteins, reported to be specific to prostate cancer and enriched in EVs - six-transmembrane epithelial antigen1 and 2 (STEAP1, STEAP2), and multidrug-resistance-associated protein 4 (ABCC4) - are S-palmitoylated [86]. Furthermore, the treatment with the palmitoylation inhibitor 2BP induced a general decrease of those three antigens in EVs. This observation confirms a role of protein palmitoylation in their trafficking and sorting to EVs [86,120]. In line with this finding, it has been shown that palmitoylated proteins accumulate also on the breast cancer-derived sEV surface and bind to Toll-like receptor 2 (TLR2), a signaling adapter in the NF-kB pathway contributing to NF-kB stimulation [121]. TLR2-mediated NF-kB activation in macrophages results in secretion of pro-inflammatory cytokines to promote cancer progression. Therefore, targeting the process of protein palmitoylation in cancer cells could inhibit the sEV-mediated NF-kB activity in macrophages and, in turn, the pro-metastatic adaptation of tumor microenvironments [121].

More in general, the possibility to target the palmitoylation pathway has to be considered as a therapeutical approach to revert EV-mediated pathological conditions. Thus, with progressively discovered information on EV-associated lipidations, may be elucidated different mechanisms in physiology as well as pathology, with implications for preclinical and clinical settings.

5. Applications of EV-related lipidation

Focusing on EV protein-related lipidation can also help to expand the EV-based biotechnology applications (Fig. 1). High-resolution quantitative MS-based proteomics, high-resolution liquid chromatography, data searching algorithms, and improved genetic engineering have provided key insights into the identification and characterization of novel EV-based protein modifications. Knowledge of these modifications and interactions will provide advances in EV-based therapeutic strategies based on how distinct cargoes are localized, sorted, and delivered [122–124].

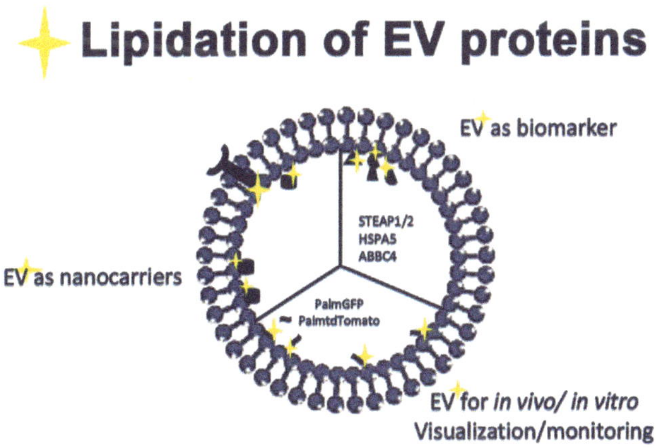

Fig. 1 Scheme of different biotech applications of EV-associated lipidation.

5.1 Functionalization of EVs through lipidation

Reprogramming native EVs by biotech approaches offers an attractive prospect for extending the therapeutic capabilities of EVs beyond their natural properties [125]. The most common strategies for loading therapeutic cargoes usually occur after EV isolation, mixing EVs with the desired cargo and subjecting them to passive incubation, electroporation, freeze-thaw cycling, sonication, extrusion, or membrane permeabilization with saponin, among various techniques. Moreover, considering the biogenetic processes of EVs, there is also the possibility that synthetic or modified biomolecules can be incorporated into EVs, by engineered the EV-producing cells. In this contest, various approaches have been studied for reprogramming EVs using protein fusion by inserting lipid anchors or functional domains. In this context, various reversible PTMs were tested and compared using lipidation motifs for myristoylation, S-palmitoylation, or prenylation, even though S-palmitoylation [126].

Palmitoylated proteins are predominantly associated with the inner membranes of EVs. Thus, this strategy is more suitable for packing proteins into EVs instead of surface display. The design and generation of fusion proteins with specific activity could be challenging for the transfection of EV-producing cells that may affect both the cargo and protein composition of EVs, and could change pathway activation at cellular levels [127].

5.2 Use of lipidation for EV visualization

Moreover, another aspect to be improved concerns the accurate visualization and monitoring of EVs, which is crucial for understanding their biological mechanisms, even if EV labeling with organic dyes may lead to inaccurate interpretations [128]. To overcome these limitations, Lai et al. developed a next-generation reporter to enable EV imaging from organisms with super-resolution, by engineering an enhanced green fluorescent protein (EGFP) fused to a palmitoylation signal, termed PalmGFP. In this work, it was developed a multi-directional imaging system of endogenously released EVs with resonance energy transfer, using lentivirally transduced cells as EV donor cells that stably expressed PalmGFP [129,130]. In particular, they genetically fused the EGFP and tandem dimer Tomato (tdTomato) in-frame to the N-terminal using a palmitoylation signal (PalmGFP, PalmtdTomato) to label the inner membrane leaflet of cells and EVs. As EVs originate from membranes, tagging the plasma membrane with fluorescent proteins would enable EV labeling [131]. Then, the high-resolution tracking of EV uptake, trough reporter strategy [129] could elucidate, for example, the spatio-temporal properties and mechanisms of cancer related to multi-directional EV exchange, during disease progression. Indeed, this live-cell imaging technology, based on confocal microscopy, permits to label multiple EV populations for tracking tumor release, using a palmitoylation signal. Moreover, it was possible to monitor EV-RNA cargoes, trough transcripts encoding PalmtdTomato that were tagged with bacteriophage MS2 RNA binding sequences and detected by the co-expression of MS2 coat protein (MS2CP) fused with EGFP. In this way, they reveal the rapid dynamics of both EV uptake and the translation of EV-delivered cargo mRNAs in cancer cells [129,131]. These studies confirm that EV-mediated communication, with the delivery of functional mRNA, is dynamic and multidirectional between cells [132]. Thus, palmitoylated fluorescent EV reporters can enable: the labeling of multiple EV types, irrespective of their biogenesis; the semi-quantification of EVs; the time-lapse live-cell imaging of EV release and uptake; the EV exchange between different cell populations.

Moreover, these reporters enabled *in vivo* observation of endogenously released EVs of implanted EL4 thymoma in C57BL/6 mice by multiphoton intravital microscopy (MP-IVM) [129,133].

5.3 EV protein lipidation as biomarkers

Lipidations associated with EVs have recently been identified as biomarkers for the diagnosis of numerous diseases, including cancer. In particular, circulating EVs as biomarkers are often related to the deregulation of protein palmitoylation, which plays a key role in the pathogenesis of many diseases. Several studies have been focusing on the identification of palmitoylated proteins in EVs through a viable liquid biopsy to identify cancer pathologies [134].

The first profile of the palmitoyl proteome of EVs was described by Mariscal et al. [86]. They developed a novel strategy for the early detection of clinically significant prostate cancer (PC) by high-throughput palmitoyl-proteomics of EVs. The palmitoyl proteome was used for marker identification of aggressive PC due to an enrichment of putative palmitoylated proteins in EVs in comparison with cells. Moreover, most of the plasma proteins that contaminate the EV preparations are not palmitoylated. For example, the palmitoylation of HSPA5 was enriched in sEVs, while the not palmitoylated HSPA5 was enriched in lEVs. Specific palmitoylated proteins, such as

STEAP1, STEAP2, and ABBC4, are enriched in both PC-derived EV populations, in comparison with the originating cell lines. Interestingly, the presence of the above proteins in EVs was significantly reduced upon the inhibition of palmitoylation in the producing cells [86,135]. This is in agreement with experiments on a group of patients with metastatic PC and helps narrow down a list of palmitoylated proteins that could be used as biomarkers in plasma. In stark contrast, the palmitoyl-proteome of EVs in patient plasma differs drastically compared to the whole plasma prior to EV purification. These results suggest that palmitoylation may be involved in the differential sorting of proteins to distinct EV populations and allow for better detection of disease biomarkers that would not be possible to identify otherwise. In addition, Zhou et al. have developed a novel method that uses a metabolic labeling-independent, cysteine-Acentric approach, namely low background ABE (LB-ABE), to study the EV palmitoyl-proteome in order to compare the palmitoyl-proteome of lEVs and sEVs versus their parental cells [136].

In conclusion, the study of the proteomic profiling associated with EV protein-related lipidation has a dual purpose: the exploitation of EVs for the relatively non-invasive collection and enrichment of disease-associated proteins to be used as prognostic and diagnostic biological markers, and the investigation of the roles of EVs and their proteomes in biological phenomena such as disease progression [134].

6. Concluding remarks and perspectives

Lipidation occurs in many EV proteins and regulates numerous biological pathways, such as protein trafficking, sorting, or signal transduction [119]. Here, we have discussed the functions of these modifications, the enzymes that control them, and the tools and technologies developed to study them. Remarkable studies have been reported regarding S-palmitoylation; it is the most common PTM of EV proteins. In particular, this post-translational lipid modification is involved in the trafficking of membrane proteins, achievement of stable protein conformations, and stabilization of protein interactions. An example is the regulation of the Alix protein. It has been reported that S-palmitoylation modulates its functions, facilitating interactions among EV-specific regulators, and maintaining the proper structural organization of EV membranes.

Nevertheless, to date the fundamental mechanisms underlying these lipid modifications are still lacking. Indeed, it is not fully clear in which cellular compartment/s protein lipidation occurs, nor whether it can also take place within EVs themselves. Lipidated proteins and several classes of palmitoyl-related enzymes are usually membrane-bound, residing in the endoplasmic reticulum, Golgi or plasma membrane. Lipidation influences the association of proteins with the plasma membrane or modulates their sorting inside EVs. It is plausible that this modification causes a conformational change in the proteins, which can also promote their polymerization; this would probably allow the packing of proteins inside EVs and protect protein folding from the acidic pH of EVs.

The reversible and dynamic regulation of lipid modification is another area that has not been thoroughly investigated. For example, can metabolism affect protein lipidation in EVs? Glycolytic enzymes are among the most frequently identified proteins in EV proteomics [137]. Accordingly, some studies have demonstrated the presence of ATP molecules, which are necessary for active reactions, in EVs [138]. Can the activation of certain pathways affect lipid modifications inside EVs? In line with the exosome database (ExoCarta), many enzymes involved in lipidation reactions have been found in EVs. In this scenario, it is possible to imagine that EV cargo should not only be trafficked to cells, but it could undergo active molecular changes that can switch on/off its functions. In particular, several proteomic studies have shown the presence of lipidic enzymes in EVs, especially those released by human cancer cells [139–141]. Among these are some palmitoyl acyltransferases (ZDHHC9, ZDHHC13, ZDHHC20), palmitoyl thioesterases (APT1, APT2, PTT2), GPI transamidases (GPI), N-myristoyltransferases (NMT1, NMT2) and farnesyltransferases (FNTA) (ExoCarta). Despite this evidence, there are no scientific reports about their effective activity in EVs.

Given the scientific community's great attention to EVs, especially for their clinical relevance and application, it would be interesting to develop new therapeutic strategies related to the lipidation of EV proteins as pharmacological targets [142,143].

In conclusion, this chapter underlined the knowledge gained and the availability of numerous tools and technologies in lipidation of EV proteins. However, it is desirable that unsolved challenges should be appropriately studied in depth, in order to broaden our knowledge in this field.

Acknowledgments

The authors acknowledge financial support from the VES4US project funded by the European Union's Horizon 2020 research and innovation program under grant agreement No 801338.

References

[1] S. Gill, R. Catchpole, P. Forterre, Extracellular membrane vesicles in the three domains of life and beyond, FEMS (Fed. Eur. Microbiol. Soc.) Microbiol. Rev. 43 (3) (2019) 273–303.
[2] G. Van Niel, G. D'Angelo, G. Raposo, Shedding light on the cell biology of extracellular vesicles, Nat. Rev. Mol. Cell Biol. 19 (4) (2018) 213–228.
[3] D.P. Romancino, G. Paterniti, Y. Campos, A. De Luca, V. Di Felice, A. d'Azzo, A. Bongiovanni, Identification and characterization of the nano-sized vesicles released by muscle cells, FEBS (Fed. Eur. Biochem. Soc.) Lett. 587 (2013) 1379–1384.
[4] J. Kowal, G. Arras, M. Colombo, M. Jouve, J.P. Morath, B. Primdal-Bengtson, F. Dingli, D. Loew, M. Tkach, C. Théry, Proteomic comparison defines novel markers to characterize heterogeneous populations of extracellular vesicle subtypes, P.N.A.S. 8 (2016) 968–977.
[5] V.R. Minciacchi, S. You, C. Spinelli, S. Morley, M. Zandian, P. Aspuria, L. Cavallini, C. Ciardiello, M.R. Sobreiro, M. Morello, et al., Large oncosomes contain distinct protein cargo and represent a separate functional class of tumor-derived extracellular vesicles, Oncotarget 6 (2015) 11327–11341.
[6] M. Mathieu, L. Martin-Jaular, G. Lavieu, C. Thery, Specificities of secretion and uptake of exosomes and other extracellular vesicles for cell-to-cell communication, Nat. Cell Biol. 21 (1) (2019) 9–17.
[7] D.M. Pegtel, S.J. Gould, Exosomes, Annu. Rev. Biochem. 88 (2019) 487–514.
[8] A.S. Venne, L. Kollipara, R.P. Zahedi, The next level of complexity: crosstalk of posttranslational modifications, Proteomics 14 (4–5) (2014) 513–524.
[9] K. Szabo-Taylor, B. Ryan, X. Osteikoetxea, T.G. Szabo, B. Sodar, M. Holub, A. Nemeth, K. Paloczi, E. Pallinger, P. Winyard, E.I. Buzas, Oxidative and other posttranslational modifications in extracellular vesicle biology, Semin. Cell Dev. Biol. 40 (2015) 8–16.
[10] O. Moreno-Gonzalo, I. Fernandez-Delgado, Sanchez-Madrid F Post-translational add- ons mark the path in exosomal protein sorting, Cell. Mol. Life Sci. 75 (1) (2018) 1–19.
[11] J.E. Smotrys, M.E. Linder, Palmitoylation of intracellular signaling proteins: regulation and function, Annu. Rev. Biochem. 73 (2004) 559–587.
[12] B. Claridge, K. Kastaniegaard, A. Stensballe, D.W. Greening, Post-translational and transcriptional dynamics regulating extracellular vesicle biology, Expert Rev. Proteomics 16 (1) (2019) 17–31.
[13] H. Ageta, K. Tsuchida, Post-translational modification and protein sorting to small extracellular vesicles including exosomes by ubiquitin and UBLs, Cell. Mol. Life Sci. 76 (24) (2019) 4829–4848.
[14] G. Triola, The protein lipidation and its analysis, J. Glycom. Lipidom. (2011) 2.
[15] C. Fankhauser, S.W. Homans, J.E. Thomasoates, et al., Structures of glycosylphosphatidylinositol membrane anchors from *Saccharomyces cerevisiae*, J. Biol. Chem. 268 (35) (1993) 26365–26374.
[16] W.L. Roberts, J.J. Myher, A. Kuksis, M.G. Low, T.L. Rosenberry, Lipid analysis of the glycoinositol phospholipid membrane anchor of human erythrocyte acetylcholinesterase Palmitoylation of inositol results in resistance to phosphatidylinositol-specific phospholipase, C. J Biol Chem. 263 (35) (1988) 18766–18775.

[17] I. Levental, M. Grzybek, K. Simons, Greasing their way: lipid modifications determine protein association with membrane rafts, Biochemistry 49 (30) (2010) 6305–6316.
[18] J.K. Nagpal, S. Dasgupta, S. Jadallah, et al., Profiling the expression pattern of GPI transamidase complex subunits in human cancer, Mod. Pathol. 21 (8) (2008) 979–991.
[19] M.D. Marmor, M. Julius, The function of GPI-anchored proteins in T cell development, activation and regulation of homeostasis, J. Biol. Regul. Homeost. Agents 14 (2) (2000) 99–115.
[20] F.L. Zhang, P.J. Casey, Protein prenylation: molecular mechanisms and functional consequences, Annu. Rev. Biochem. 65 (1996) 241–269.
[21] M.N. Ashby, X. Caa, Converting enzymes, Curr. Opin. Lipidol. 9 (2) (1998) 99–102.
[22] M.D. Resh, Trafficking and signaling by fatty-acylated and prenylated proteins, Nat. Chem. Biol. 2 (11) (2006) 584–590.
[23] A. Hentschel, R.P. Zahedi, R. Ahrends, Protein lipid modifications–More than just a greasy ballast, Proteomics 16 (5) (2016) 759–782.
[24] T.E. Smeland, M.C. Seabra, J.L. Goldstein, M.S. Brown, Geranylgeranylated Rab proteins terminating in Cys-Ala-Cys, but not Cys-Cys, are carboxyl-methylated by bovine brain membranes in vitro, Proc. Natl. Acad. Sci. U.S.A. 91 (22) (1994) 10712–10716.
[25] J.H. Jackson, C.G. Cochrane, J.R. Bourne, P.A. Solski, J.E. Buss, C.J. Der, Farnesol modification of Kirsten-ras exon 4B protein is essential for transformation, Proc. Natl. Acad. Sci. U.S.A. 87 (8) (1990) 3042–3046.
[26] D.A. Towler, S.P. Adams, S.R. Eubanks, et al., Purification and characterization of yeast myristoyl CoA:protein N-myristoyltransferase, Proc. Natl. Acad. Sci. U.S.A. 84 (9) (1987) 2708–2712.
[27] D.A. Towler, S.R. Eubanks, D.S. Towery, S.P. Adams, L. Glaser, Amino-terminal processing of proteins by N-myristoylation. Substrate specificity of N-myristoyl transferase, J. Biol. Chem. 262 (3) (1987) 1030–1036.
[28] G.L. Vilas, M.M. Corvi, G.J. Plummer, A.M. Seime, G.R. Lambkin, L.G. Berthiaume, Posttranslational myristoylation of caspase-activated p21-activated protein kinase 2 (PAK2) potentiates late apoptotic events, Proc. Natl. Acad. Sci. U.S.A. 103 (17) (2006) 6542–6547.
[29] J. Zha, S. Weiler, K.J., Oh, M.C. Wei, S.J. Korsmeyer, Posttranslational N-myristoylation of BID as a molecular switch for targeting mitochondria and apoptosis, Science 290 (5497) (2000) 1761–1765.
[30] D.D. Martin, E. Beauchamp, L.G. Berthiaume, Post-translational myristoylation: fat matters in cellular life and death, Biochimie 93 (1) (2011) 18–31.
[31] S. Mclaughlin, A. Aderem, The myristoyl-electrostatic switch: a modulator of reversible protein-membrane interactions, Trends Biochem. Sci. 20 (7) (1995) 272–276.
[32] J. Goldberg, Structural basis for activation of ARF GTPase: mechanisms of guanine nucleotide exchange and GTP-myristoyl switching, Cell 95 (2) (1998) 237–248.
[33] P. Mill, A.W. Lee, Y. Fukata, et al., Palmitoylation regulates epidermal homeostasis and hair follicle differentiation, PLoS Genet. 5 (11) (2009) e1000748.
[34] H. Gerlach, V. Laumann, S. Martens, C.F. Becker, R.S. Goody, M. Geyer, HIV-1 Nef membrane association depends on charge, curvature, composition and sequence, Nat. Chem. Biol. 6 (1) (2010) 46–53.
[35] S. Maurer-stroh, F. Eisenhaber, Myristoylation of viral and bacterial proteins, Trends Microbiol. 12 (4) (2004) 178–185.

[36] T.M. Hall, J.A. Porter, K.E. Young, E.V. Koonin, P.A. Beachy, D. J Leahy, Crystal structure of a Hedgehog autoprocessing domain: homology between Hedgehog and self-splicing proteins, Cell 91 (1) (1997) 85−97.

[37] J.A. Porter, D.P. Von kessler, S.C. Ekker, et al., The product of hedgehog autoproteolytic cleavage active in local and long-range signalling, Nature 374 (6520) (1995) 363−366.

[38] K.E. Ryan, C. Chiang, Hedgehog secretion and signal transduction in vertebrates, J. Biol. Chem. 287 (22) (2012) 17905−17913.

[39] J.A. Porter, S.C. Ekker, W.J. Park, et al., Hedgehog patterning activity: role of a lipophilic modification mediated by the carboxy-terminal autoprocessing domain, Cell 86 (1) (1996) 21−34.

[40] J. Briscoe, Y. Chen, T.M. Jessell, G. Struhl, A hedgehog-insensitive form of patched provides evidence for direct long-range morphogen activity of sonic hedgehog in the neural tube, Mol. Cell. 7 (6) (2001) 1279−1291.

[41] D. Panáková, H. Sprong, E. Marois, C. Thiele, S. Eatonm, Lipoprotein particles are required for Hedgehog and Wingless signalling, Nature 435 (7038) (2005) 58−65.

[42] F.T. Stevenson, S.L. Bursten, R.M. Locksley, D.H. Lovett, Myristyl acylation of the tumor necrosis factor alpha precursor on specific lysine residues, J. Exp. Med. 176 (4) (1992) 1053−1062.

[43] F.T. Stevenson, S.L. Bursten, C. Fanton, R.M. Locksley, D.H. Lovett, The 31-kDa precursor of interleukin 1 alpha is myristoylated on specific lysines within the 16-kDa N-terminal propiece, Proc. Natl. Acad. Sci. U.S.A. 90 (15) (1993) 7245−7249.

[44] K.L. Schey, D.B. Gutierrez, Z. Wang, J. Wei, A.C. Grey, Novel fatty acid acylation of lens integral membrane protein aquaporin-0, Biochemistry v49 (45) (2010) 9858−9865.

[45] T.C. Turek-etienne, C.L. Strickland, M.D. Distefano, Biochemical and structural studies with prenyl diphosphate analogues provide insights into isoprenoid recognition by protein farnesyl transferase, Biochemistry 42 (13) (2003) 3716−3724.

[46] X. Zhang, N. A Spiegelman, O. D Nelson, H. Jing, H. Lin, SIRT6 regulates Ras-related protein R-Ras2 by lysine defatty-acylation, Elife (2017) 6.

[47] C.T. Lim, B. Kola, M. Korbonits, The ghrelin/GOAT/GHS-R system and energy metabolism, Rev. Endocr. Metab. Disord. 12 (3) (2011) 173−186.

[48] M.S. Taylor, T.R. Ruch, P.Y. Hsiao, et al., Architectural organization of the metabolic regulatory enzyme ghrelin O-acyltransferase, J. Biol. Chem. 288 (45) (2013) 32211−32228.

[49] M.D. Resh, Palmitoylation of ligands, receptors, and intracellular signaling molecules, Sci. STKE 2006 (359) (2006) re14.

[50] C. Kleuss, E. Krause, Galph(s)a palmitoylated is at the N-terminal glycine, EMBO J. 22 (4) (2003) 826−832.

[51] J.A. Buglino, M. D Resh, Hhat is a palmitoylacyltransferase with specificity for N-palmitoylation of Sonic Hedgehog, J. Biol. Chem. 283 (32) (2008) 22076−22088.

[52] R.C. Aicart, R.A. Valero, C.I. Rodriguez, Protein palmitoylation and subcellular trafficking, Biochim. Biophys. Acta 1808 (12) (2011) 2981−2994.

[53] C. Salaun, J. Greaves, L.H. Chamberlain, The intracellular dynamic of protein palmitoylation, J. Cell Biol. 191 (7) (2010) 1229−1238.

[54] M.D. Resh, Fatty acylation of proteins: the long and the short of it, Prog. Lipid Res. 63 (2016) 120−131.

[55] J. Greaves, L.H. Chamberlain, DHHC palmitoyl transferases: substrate interactions and (patho)physiology, Trends Biochem. Sci. 36 (5) (2011) 245−253.

[56] B.C. Jennings, M.E. Linder, DHHC protein S-acyltransferases use similar ping-pong kinetic mechanisms but display different acyl-CoA specificities, J. Biol. Chem. 287 (10) (2012) 7236−7245.

[57] R. Zeidman, C.S. Jackson, A.I. Magee, Protein acyl thioesterases, (Review), Mol. Membr. Biol. 26 (1) (2009) 32–41.
[58] J. Rios-esteves, B. Haugen, M.D. Resh, Identification of key residues and regions important for porcupine-mediated Wnt acylation, J. Biol. Chem. 289 (24) (2014) 17009–17019.
[59] C.Y. Janda, D. Waghray, A.M. Levin, C. Thomas, K.C. Garcia, Structural basis of Wnt recognition by frizzled, Science 337 (6090) (2012) 59–64.
[60] G.S. Coombs, J. Yu, C.A. Canning, et al., WLS-dependent secretion of WNT3A requires Ser209 acylation and vacuolar acidification, J. Cell Sci. 123 (Pt 19) (2010) 3357–3367.
[61] S. Kakugawa, P.F. Langton, M. Zebisch, et al., Notum deacylates Wnt proteins to suppress signalling activity, Nature 519 (7542) (2015) 187–192.
[62] Z. Kelman, V. Naktinis, M. O'Donnell, Radiolabeling of proteins for biochemical studies, Methods Enzymol. 262 (1995) 430–442.
[63] N. Sakurai, T. Utsumi, Posttranslational N-myristoylation is required for the anti-apoptotic activity of human tGelsolin, the C-terminal caspase cleavage product of human gelsolin, J. Biol. Chem. 281 (2006) 14288–14295.
[64] T. Utsumi, N. Sakura, K. Nakano, R. Ishisaka, C-terminal 15 kDa fragment of cytoskeletal actin is posttranslationally N-myristoylated upon caspase- mediated cleavage and targeted to mitochondria, FEBS Lett. 539 (2003) 37–44.
[65] S. Kjellstrom, O.N. Jensen, In situ liquid - liquid extraction as a sample preparation method for matrix-assisted laser desorption/ionization MS analysis of polypeptide mixtures, Anal. Chem. 75 (2003) 2362–2369.
[66] E. Sachon, P.F. Nielsen, O.N. Jensen, Characterization of N-palmitoylated human growth hormone by in situ liquid-liquid extraction and MALDI tandem mass spectrometry, J. Mass Spectrom. 42 (2007) 724–734.
[67] M.P. Washburn, D. Wolters, J.R. Yates, Large-scale analysis of the yeast proteome by multidimensional protein identification technology, Nat. Biotechnol. 19 (2001) 242–247.
[68] M. T Forrester, D.T. Hess, J.W. Thompson, R. Hultman, M.A. Moseley, et al., Site-specific analysis of protein S-acylation by resin-assisted capture, J. Lipid Res. 52 (2011) 393–398.
[69] M.T. Forrester, J.W. Thompson, M.W. Foster, L. Nogueira, M.A. Moseley, et al., Proteomic analysis of S-nitrosylation and denitrosylation by resinassisted capture, Nat. Biotechnol. 27 (2009) 557–559.
[70] N. Sorek, S. Yalovsky, Analysis of protein S-acylation by gas chromatography-coupled mass spectrometry using purified proteins, Nat. Protoc. 5 (2010) 834–840.
[71] R.C. Drisdel, W.N. Green, Labeling and quantifying sites of protein palmitoylation, Biotechniques 36 (2004) 276–285.
[72] G.S. Brigidi, S.X. Bamji, Detection of protein palmitoylation in cultured hippocampal neurons by immunoprecipitation and acyl-biotin exchange (ABE), JoVE (72) (2013) 50031.
[73] D.P. Romancino, V. Buffa, S. Caruso, I. Ferrara, S. Raccosta, A. Notaro, Y. Campos, R. Noto, V. Martorana, A. Cupane, A. Giallongo, A. d'Azzo, M. Manno, A. Bongiovanni, Palmitoylation is a post-translational modification of Alix regulating the membrane organization of exosome-like small extracellular vesicles, Biochim. Biophys. Acta Gen. Subj. 1862 (2018) 2879–2887.
[74] M.A. Kostiuk, M.M. Corvi, B.O. Keller, P. Glummer, J.A. Prescher, et al., Identification of palmitoylated mitochondrial proteins using a bio-orthogonal azido-palmitate analogue, Faseb. J. 22 (2008) 721–732.
[75] B.R. Martin, B.F. Cravatt, Large-scale profiling of protein palmitoylation in mammalian cells, Nat. Methods 6 (2009) 135–138.

[76] J.S. Yount, G. Charron, H.C. Hang, Bioorthogonal proteomics of 15-hexadecynyloxyacetic acid chemical reporter reveals preferential targeting of fatty acid modified proteins and biosynthetic enzymes, Bioorg. Med. Chem. 20 (2012) 650–654.
[77] E.M. Sletten, C.R. Bertozzi, Bioorthogonal chemistry: fishing for selectivity in a sea of functionality, Angew Chem. Int. Ed. Engl. 48 (2009) 6974–6998.
[78] Y. Kho, S.C. Kim, C. Jiang, D. Barma, S.W. Kwon, et al., A tagging-viasubstrate technology for detection and proteomics of farnesylated proteins, Proc. Natl. Acad. Sci. U.S.A. 101 (2004) 12479–12484.
[79] E. Saxon, C.R. Bertozzi, Cell surface engineering by a modified Staudinger reaction, Science 287 (2000) 2007–2010.
[80] U.T. Nguyen, Z. Guo, C. Delon, Y. Wu, C. Deraeve, et al., Analysis of the eukaryotic prenylome by isoprenoid affinity tagging, Nat. Chem. Biol. 5 (2009) 227–235.
[81] A.F. Berry, W.P. Heal, A.K. Tarafder, T. Tolmachova, R.A. Baron, et al., Rapid multilabel detection of geranylgeranylated proteins by using bioorthogonal ligation chemistry, Chembiochem 11 (2010) 771–773.
[82] B.R. Martin, B.F. Cravatt, Global profiling of dynamic protein palmitoylation, Nat. Methods 9 (1) (2011) 84–89, 6.
[83] F. Wolbers, P. ter Braak, S. Le Gac, R. Luttge, H. Andersson, et al., Viability study of HL60 cells in contact with commonly used microchip materials, Electrophoresis 27 (2006) 5073–5080.
[84] A.E. Speers, G.C. Adam, B.F. Cravatt, Activity-based protein profiling in vivo using a copper(I)-catalyzed azide-alkyne [3+2] cycloaddition, J. Am. Chem. Soc. 125 (2003) 4686–4687.
[85] P.V. Chang, J.A. Prescher, E.M. Sletten, J.M. Baskin, I.A. Miller, et al., Copper-free click chemistry in living animals, Proc. Natl. Acad. Sci. U.S.A. 107 (2010) 1821–1826.
[86] J. Mariscal, T. Vagner, M. Kim, B. Zhou, A. Chin, M. Zandian, M.R. Freeman, S. You, A. Zijlstra, W. Yang, D. Di Vizio, Comprehensive palmitoyl-proteomic analysis identifies distinct protein signatures for large and small cancer-derived extracellular vesicles, bioRxiv (2019) 787499, https://doi.org/10.1101/787499.
[87] M. Wang, S. Altinoglu, Y.S. Takeda, Q. Xu, Integrating protein engineering and bioorthogonal click conjugation for extracellular vesicle modulation and intracellular delivery, PloS One 10 (11) (2015) e0141860, 2015 Nov 3.
[88] S. Chen, S.C. Shiesh, G.B. Lee, C. Chen, Two-step magnetic bead-based (2MBB) techniques for immunocapture of extracellular vesicles and quantification of microRNAs for cardiovascular diseases: a pilot study, PloS One 15 (2) (2020) e0229610.
[89] A. Forterre, A. Jalabert, E. Berger, et al., Proteomic analysis of C2C12 myoblast and myotube exosome-like vesicles: a new paradigm for myoblast-myotube cross talk? PloS One 9 (1) (2014) e84153.
[90] M.S. Panagopoulou, A.W. Wark, D.J.S. Birch, C.D. Gregory, Phenotypic analysis of extracellular vesicles: a review on the applications of fluorescence, J. Extracell. Vesicles 9 (1) (2020) 1710020.
[91] S. Maurer-Stroh, et al., N-terminal N-myristoylation of proteins: prediction of substrate proteins from amino acid sequence, J. Mol. Biol. 317 (2002) 541–557.
[92] G. Bologna, et al., N-terminal myristoylation predictions by ensembles of neural networks, Proteomics 4 (2004) 1626–1632.
[93] S. Maurer-Stroh, et al., Refinement and prediction of protein prenylation motifs, Genome Biol. 6 (2005) 1.
[94] J. Ren, et al., CSS-Palm 2.0: an updated software for palmitoylation sites prediction, Protein Eng. Des. Sel. 21 (2008) 639–644.
[95] X.-B. Wang, et al., Prediction of palmitoylation sites using the composition of k-spaced amino acid pairs, Protein Eng. Des. Sel. 22 (2009) 707–712.

[96] M. Blanc, et al., SwissPalm: protein palmitoylation database, F1000Res 4 (2015).
[97] S. Li, et al., In silico identification of protein S-palmitoylation sites and their involvement in human inherited disease, J. Chem. Inf. Model. 55 (2015) 2015–2025.
[98] W. Ning, P. Jiang, Y. Guo, C. Wang, X.,W. Tan Zhang, D. Peng, Y. Xue, GPS-Palm: a deep learning-based graphic presentation system for the prediction of S-palmitoylation sites in proteins, Bioinformation (2020) bbaa038, https://doi.org/10.1093/bib/bbaa038 [published online ahead of print, 2020 Apr 3].
[99] Y. Xie, Y. Zheng, H. Li, et al., GPS-Lipid: a robust tool for the prediction of multiple lipid modification sites, Sci. Rep. 6 (2016) 28249, https://doi.org/10.1038/srep28249.
[100] L.R. Fernandes, V.B. Rocha, V.C. Carregari, A. Urbani, G. Palmisano, A perspective on extracellular vesicles proteomics, Front Chem 5 (2017) 102.
[101] D.W. Greening, R.J. Simpson, Understanding extracellular vesicle diversity - current status, Expert Rev. Proteomics 15 (11) (2018) 887–910.
[102] M. Blanc, F. David, L. Abrami, D. Migliozzi, F. Armand, J. Bürgi, et al., SwissPalm: protein palmitoylation database, F1000Res 4 (2015) 261.
[103] B. Tancini, S. Buratta, K. Sagini, et al., Insight into the role of extracellular vesicles in lysosomal storage disorders, Genes 10 (7) (2019).
[104] K.B. Beer, A.M. Wehman, Mechanisms and functions of extracellular vesicle release in vivo-What we can learn from flies and worms, Cell Adhes. Migrat. 11 (2) (2017) 135–150.
[105] S. Liégeois, A. Benedetto, J.M. Garnier, Y. Schwab, M. Labouesse, The V0-ATPase mediates apical secretion of exosomes containing Hedgehog-related proteins in *Caenorhabditis elegans*, J. Cell Biol. 173 (6) (2006) 949–961.
[106] K. Roy, E.P. Marin, Lipid modifications in cilia biology, J. Clin. Med. 8 (7) (2019).
[107] S. Takada, S. Fujimori, T. Shinozuka, R. Takada, Y. Mii, Differences in the secretion and transport of Wnt proteins, J. Biochem. 161 (1) (2017) 1–7.
[108] E. Eitan, R.S. Petralia, Y.X. Wang, F.E. Indig, M.P. Mattson, P.J. Yao, Probing extracellular Sonic hedgehog in neurons, Biol Open 5 (8) (2016) 1086–1092.
[109] T. Shibata, J. Hadano, D. Kawasaki, X. Dong, S. Kawabata, *Drosophila* TG-A transglutaminase is secreted via an unconventional Golgi-independent mechanism involving exosomes and two types of fatty acylations, J. Biol. Chem. 292 (2017) 10723–10734.
[110] D. Perez-Hernandez, C. Gutiérrez-Vázquez, I. Jorge, S. López-Martín, A. Ursa, F. Sánchez Madrid, J. Vázquez, M. Yañez-Mó, The intracellular interactome of tetraspanin-enriched microdomains reveals their function as sorting machineries toward exosome, J. Biol. Chem. 288 (2013) 11649–11661.
[111] Z. Andreu, M. Yáñez-mó, Tetraspanins in extracellular vesicle formation and function, Front. Immunol. 5 (2014) 442.
[112] D.C. Bittel, K. Jaiswal, Contribution of extracellular vesicles in rebuilding injured muscles, Front. Physiol. 10 (2019) 828, 1-20.
[113] A.J. Jimenez, P. Maiuri, J. Lafaurie-Janvore, S. Divoux, M. Piel, F. Perez, ESCRT machinery is required for plasma membrane repair, Science 343 (2014) 1247136.
[114] L.L. Scheffer, S.C. Sreetama, N. Sharma, S. Medikayala, K.J. Brown, A. Defour, et al., Mechanism of Ca(2)(+)-triggered ESCRT assembly and regulation of cell membrane repair, Nat. Commun. 5 (2014) 5646.
[115] T. Hornemann, Palmitoylation and depalmitoylation defects, J. Inherit. Metab. Dis. 38 (1) (2015) 179–186.
[116] C. Goettsch, J.D. Hutcheson, M. Aikawa, et al., Sortilin mediates vascular calcification via its recruitment into extracellular vesicles, J. Clin. Invest. 126 (4) (2016) 1323–1336.

[117] S. Itoh, K. Mizuno, M. Aikawa, E. Aikawa, Dimerization of sortilin regulates its trafficking to extracellular vesicles, J. Biol. Chem. 293 (12) (2018) 4532–4544.
[118] Y. Zhang, X. Wu, W. Andy tao, Characterization and applications of extracellular vesicle proteome with post-translational modifications, Trends Anal. Chem. 107 (2018) 21–30.
[119] H. Jiang, X. Zhang, X. Chen, P. Aramsangtienchai, Z. Tong, H. Lin, Protein lipidation: occurrence, mechanisms, biological functions, and enabling technologiesm, Chem. Rev. 118 (3) (2018) 919–988.
[120] T. Vagner, A. Chin, J. Mariscal, S. Bannykh, D.M. Engman, D. Di vizio, Protein composition reflects extracellular vesicle heterogeneity, Proteomics 19 (8) (2019) e1800167.
[121] A. Chow, W. Zhou, L. Liu, et al., Macrophage immunomodulation by breast cancer-derived exosomes requires Toll-like receptor 2-mediated activation of NF-κB, Sci. Rep. 4 (2014) 5750.
[122] Y.V. Karpievitch, A.D. Polpitiya, G.A. Anderson, R.D. Smith, A.R. Dabney, Liquid chromatography mass spectrometry-based proteomics: biological and technological aspects, Ann. Appl. Stat. 4 (4) (2010) 1797–1823.
[123] R. Matthiesen, J. Bunkenborg, Introduction to mass spectrometry-based proteomics, Methods Mol. Biol. 1007 (2013) 1–45.
[124] B. György, M.E. Hung, X.O. Breakefield, J.N. Leonard, Therapeutic applications of extracellular vesicles: clinical promise and open questions, Annu. Rev. Pharmacol. Toxicol. 55 (2015) 439–464.
[125] E. Geeurickx, J. Tulkens, B. Dhondt, et al., The generation and use of recombinant extracellular vesicles as biological reference material, Nat. Commun. 10 (1) (2019) 3288.
[126] X. Shi, Q. Cheng, Y. Zhang, Reprogramming extracellular vesicles with engineered proteins, Methods 177 (2020) 95–102, https://doi.org/10.1016/j.ymeth.2019.09.017.
[127] R.N.P. Rodenburg, J. Snijder, M. Van De Waterbeemd, et al., Stochastic palmitoylation of accessible cysteines in membrane proteins revealed by native mass spectrometry, Nat. Commun. 8 (1) (2017) 1280.
[128] A.Y. Wu, K. Ueda, C.P. Lai, Proteomic analysis of extracellular vesicles for cancer diagnostics, Proteomics 19 (1–2) (2019) e1800162.
[129] C.P. Lai, E.Y. Kim, C.E. Badr, et al., Visualization and tracking of tumour extracellular vesicle delivery and RNA translation using multiplexed reporters, Nat. Commun. 6 (2015) 7029.
[130] C.P. Lai, O. Mardini, M. Ericsson, et al., Dynamic biodistribution of extracellular vesicles in vivo using a multimodal imaging reporter, ACS Nano 8 (1) (2014) 483–494.
[131] D.A. Zacharias, J.D. Violin, A.C. Newton, R.Y. Tsien, Partitioning of lipid-modified monomeric GFPs into membrane microdomains of live cells, Science 296 (5569) (2002) 913–916.
[132] J.A. Menard, H.C. Christianson, P. Kucharzewska, et al., Metastasis stimulation by hypoxia and acidosis-induced extracellular lipid uptake is mediated by proteoglycan-dependent endocytosis, Canc. Res. 76 (16) (2016) 4828–4840.
[133] S.T. Chuo, J.C. Chien, C.P. Lai, Imaging extracellular vesicles: current and emerging methods, J. Biomed. Sci. 25 (1) (2018) 91.
[134] I. Nazarenko, Extracellular vesicles: recent developments in technology and perspectives for cancer liquid biopsy, in: F. Schaffner, J.L. Merlin, N. von Bubnoff (Eds.), Tumor Liquid Biopsies. Recent Results in Cancer Research, vol. 215, Springer, Cham, 2020.

[135] W. Yang, D. Di vizio, M. Kirchner, H. Steen, M.R. Freeman, Proteome scale characterization of human S-acylated proteins in lipid raft-enriched and non-raft membranes, Mol. Cell. Proteomics 9 (1) (2010) 54–70.
[136] B. Zhou, Y. Wang, Y. Yan, et al., Low-background acyl-biotinyl exchange largely eliminates the coisolation of non–acylated proteins and enables deep -acylproteomic analysis, Anal. Chem. 91 (15) (2019) 9858–9866.
[137] K. Göran ronquist, Extracellular vesicles and energy metabolism, Clin. Chim. Acta 488 (2019) 116–121.
[138] K.G. Ronquist, C. Sanchez, L. Dubois, et al., Energy-requiring uptake of prostasomes and PC3 cell-derived exosomes into non-malignant and malignant cells, J. Extracell. Vesicles 5 (2016) 29877.
[139] M. He, H. Qin, T.C. Poon, et al., Hepatocellular carcinoma-derived exosomes promote motility of immortalized hepatocyte through transfer of oncogenic proteins and RNAs, Carcinogenesis 36 (9) (2015) 1008–1018.
[140] M.D. beckler, J.N. Higginbotham, J.L. Franklin, et al., Proteomic analysis of exosomes from mutant KRAS colon cancer cells identifies intercellular transfer of mutant KRAS, Mol. Cell. Proteomics 12 (2) (2013) 343–355.
[141] B. Liang, P. Peng, S. Chen, et al., Characterization and proteomic analysis of ovarian cancer-derived exosomes, J Proteomics 80 (2013) 171–182.
[142] B. Chen, Y. Sun, J. Niu, G.K. Jarugumilli, X. Wu, Protein lipidation in cell signaling and diseases: function, regulation, and therapeutic opportunities, Cell Chem Biol 25 (7) (2018) 817–831.
[143] S.A. Kooijmans, P. Vader, S.M. Van dommelen, W.W. Van Solinge, R.M. Schiffelers, Exosome mimetics: a novel class of drug delivery systems, Int. J. Nanomed. 7 (2012) 1525–1541.
[144] B. Shen, N. Wu, J.M. Yang, S.J. Gould, Protein targeting to exosomes/microvesicles by plasma membrane anchors, J. Biol. Chem. 286 (2011) 14383–14395.
[145] S. Kita, N. Maeda, I. Shimomura, Interorgan communication by exosomes, adipose tissue, and adiponectin in metabolic syndrome, J. Clin. Invest. 129 (10) (2019) 4041–4049.
[146] S.C. López, S.C. Campos, M.V. Gómez, Glycosyl-phosphatidyl-inositol (GPI)-Anchors and metalloproteases: their roles in the regulation of exosome composition and NKG2D-mediated immune recognition, Front Cell Dev Biol 4 (2016) 97.
[147] J. Zhang, Y. Zhang, C. Luo, Y. Xia, H. Chen, X. Wu, Glycosyl-phosphatidylinositol-anchored interleukin-2 expressed on tumor-derived exosomes induces antitumor immune response in vitro, Tumori 96 (3) (2010) 452–459.

CHAPTER FIVE

Pursuing mechanisms of extracellular vesicle formation. Effects of sample processing

Darja Božič[a,b,*], Matej Hočevar[c], Veno Kononenko[d], Marko Jeran[a,b], Urška Štibler[b], Immacolata Fiume[e], Manca Pajnič[a], Ljubiša Pađen[a], Ksenija Kogej[f], Damjana Drobne[d], Aleš Iglič[b,e,g], Gabriella Pocsfalvi[e] and Veronika Kralj-Iglič[a,e]

[a]Laboratory of Clinical Biophysics, Faculty of Health Sciences, University of Ljubljana, Ljubljana, Slovenia
[b]Laboratory of Physics, Faculty of Electrical Engineering, University of Ljubljana, Ljubljana, Slovenia
[c]Department of Physics and Chemistry of Materials, Institute of Metals and Technology, Ljubljana, Slovenia
[d]Department of Biology, Biotechnical Faculty, University of Ljubljana, Ljubljana, Slovenia
[e]Extracellular Vesicles and Mass Spectrometry Laboratory, Institute of Biosciences and BioResources (IBBR), National Research Council of Italy (CNR), Naples, Italy
[f]Department of Chemistry and Biochemistry, Faculty of Chemistry and Chemical Technology, University of Ljubljana, Ljubljana, Slovenia
[g]Laboratory of Clinical Biophysics, Faculty of Medicine, University of Ljubljana, Ljubljana, Slovenia
*Corresponding author: E-mail: Darja.Bozic@fe.uni-lj.si.

Contents

Advances in Biomembranes and Lipid Self-Assembly, Volume 32
ISSN 2451-9634
https://doi.org/10.1016/bs.abl.2020.09.003

Abstract

Although the role of extracellular vesicles in inter-cell communication is acknowledged, their analytics is yet problematic. Poor understanding of effects of sample processing may be one of the cornerstones of this issue. We investigated the effects of centrifugation during the vesicle isolation on the number, size, content and shape of particles in the isolates. Particles in the range of about 100–500 nm as observed by the scanning electron microscope exhibited characteristic shapes of vesicles without internal structure. Enhanced centrifugal pull resulted in increased amount of certain types of proteins in the isolates. Majority of present proteins were highly abundant human plasma proteins, including albumins and immunoglobulins. Results are in favor of the hypothesis that the isolated particles reflect the substance from which they originate (cells and the surrounding solution) and the performed processing; and therefore, vesicles in isolates should be viewed upon as a dynamic material with transient identity.

1. Introduction

Nano-to micro-sized membrane enclosed particles (vesicles) can be found in biological fluids and cell-culture media. Initially, they were called "cell dust" [1] as they were presumed to be packages of cell waste and/or a side product of cell degradation. It was later observed that they intervene in various biological processes and play important roles in development and functioning of an organism [2]. Different mechanisms of their formation were proposed, and a term extracellular vesicles (EV) was introduced to comprise the heterogeneous cluster of vesicles that can be found in biological samples.

EVs have cell-originated membrane and contain important biological molecules (proteins, lipids, nucleic acids and metabolites). Therefore, EVs are carriers of valuable information about the cells they come from, reflecting their state and condition. The foreseen potential of EVs to be used as a convenient diagnostic and potentially therapeutic tool opened an era of extensive research. However, up to day, EV analytics has not reached the level of reliability required for clinical applications. Thus, EV exploitation remains limited [3–8].

One of the bottlenecks of the progress appears to be EV harvesting, their isolation and purification [9–13]. Despite strong initiatives for standardization of these procedures [14–16], the studies of EVs remain largely incoherent. The reproducibility of the isolation experiments and also the reliability of the following characterization and functional data is deficient. The need of improvement of the existing protocols pushes the researchers into further and further changes of parameters and development of completely new approaches. Due to the high variety of methodologies, the outcomes of different studies are often hard to compare. But since the reliable and reproducible method for EV isolation is still lacking, and we do not have enough knowledge of how individual parameters influence the isolation, the standardization of protocols seems to be yet premature.

Events taking place during harvesting are still poorly understood, and processing of samples before and during the isolation procedure can cause significant changes of the EVs in the isolates [17]. Genuineness of EVs after isolation and purification procedures may be questionable and represent a big problem in the EV field. Have we crossed from disregard of the "cell dust" to some level of over-/miss-price of the microparticles' significance in the severely processed samples? We think, that in order to improve EVs' applicability, their stability must be more closely considered, primarily regarding rigidity/fluidity of the membranes and their remodeling.

The self-assembly of a biological membrane is based on the equilibrium of its components, which is on one hand directed by the cell itself (through production and regulation of membrane components) and on the other, influenced by the external factors such as chemicals, temperature, and pressure. When the constraints shaping the membrane assembly are changed, its components face a possibility (/necessity) to rearrange according to new conditions of the system. By the physical laws, the new arrangement represents the most easily accessible free energy minimum.

The membrane architecture varies from cell to cell in regard to its functionality and physiological state. The flexibility of a membrane-perimembrane construct conditions its sensitivity and response to perturbations. Disruptions may be followed by membrane spontaneous reformulation (resealing, vesicle formation) or degradation (collapse or dissociation of components).

The most common method for EV isolation from liquid samples is differential centrifugation (DC) [10,18]. Centrifuge tubes containing samples are placed in a centrifuge, where spinning creates a centrifugal field in the system that pulls the freely floating particles in the samples away from the centrifuge rotor axis. Separation is based on distinctive sedimentation coefficients (*S*) of the sample components. Sedimentation coefficients depend on particles' mass (m), radius (r) and on viscosity of the medium (η), according to relation $S = m/6\pi r\eta$. Differential centrifugation (DC) is therefore a centrifugation method in which different species are selectively spun down from a mixture by a series of increasing centrifugal forces. General pitfalls of DC include variability in sedimentation efficiencies [19,20], accompanied with co-sedimentation of undesired material (proteins and lipoproteins [21–23], RNA aggregates [24], viral particles) with similar S value; and EV aggregation and/or disruption [12,25,26]. However, in spite of vast use of ultracentrifugation-based processes, the knowledge on the effect of high centrifugal force and the concentrating of EVs throughout the sample processing is yet rudimentary.

Due to their small size, a prolonged action (hours) of a strong force (high speed centrifugation, ultracentrifugation, UC) is required to sediment EVs. The time consumption of the procedure paralyzes the high-strength studies with large number of samples, and makes EV centrifugation inconvenient for routine use in clinical practice. Alternative isolation techniques were recently proposed to overcome these issues. Ultrafiltration, flow field-flow fractionation, dialysis, size exclusion chromatography, microchip-based techniques and precipitation-based methods are being developed, and used alone or in combination with ultracentrifugation-based methods. Immunoaffinity-based isolation can also be applied to harvest EVs with particular surface protein composition [27–29]. Recently, commercial kits were launched, e.g. ExoQuick (System Biosciences) and Total Exosome Isolation kits (TEI, Invitrogen) based on polymer precipitation, qEV (Izon) based on size exclusion chromatography, (UF, Millipore) based on ultrafiltration, and exoEasy (Qiagen) based on membrane-based affinity binding [30]. However, different isolation methods were found to lead to different EV populations and different level of purity of the EV isolates [31]. Furthermore, damage or alteration to the EV structure may occur during the harvesting and assessment procedures, due to mechanical/thermal stress and chemical reactions. Although the suggested methods are faster and easier to apply than DC, a recent thorough comparison between isolates that were obtained by various methods showed that among the tested, ultracentrifugation-based methods are still preferential as regards the isolate purity [30]. Also, with available centrifuge and high throughput of samples, this technique may prove cost-effective and environment-liable.

To improve the repeatability and accuracy of the isolation procedure we examined the effects of centrifugation on biological material. In order to reveal what happens with the plasma components that are not successfully

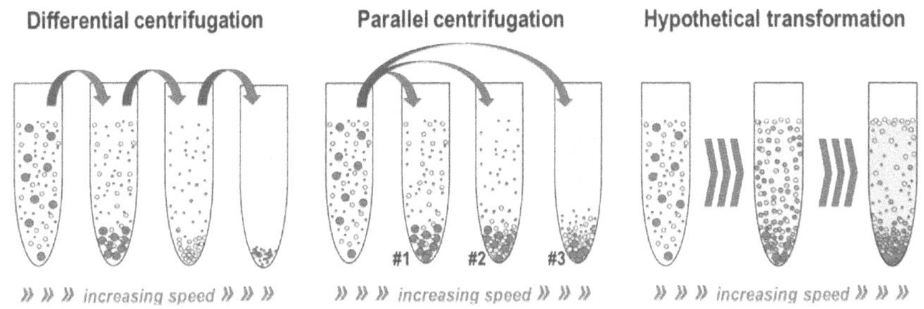

Fig. 1 Schematic representation of centrifugation approaches. Differential centrifugation is a common method for separation of particles with different sedimentation coefficients. Larger and denser particles have higher sedimentation rate, and therefore concentrate in the pellet faster than smaller and thinner ones. Low speed centrifugation is therefore used to remove cells and larger particles, while supernatant is subjected to higher speed centrifugation to sediment smaller particles. The parallel centrifugation was performed in this study to test the hypothesis of sample transformation during harvesting, and investigate possibilities of tearing of particles and reorganization of the present components upon centrifugation processing.

removed in the appropriate steps of DC, blood plasma was subjected to parallel centrifugation (schematic comparison of centrifugation-based approaches is represented in Fig. 1). Different centrifugation accelerations were tested, starting from lower (i.e. 10 000 g), and ending with 100 000 g, which is commonly used to harvest small EVs (exosomes) [10]. A protocol that is frequently used for clinical studies [17,32–46] was also inspected. We closely followed the transformation of the material from mild to the keenest conditions. Flow cytometry (FCM) and dynamic light scattering (DLS) were employed for particle quantification and size characterization. Attention was payed to the effects of freezing/thawing of the starting sample, and UC-processing temperature on the resulting isolates. Gel-based proteomics was performed on a set of isolates to reveal their protein content and composition. Morphology of isolates was examined by scanning electron microscopy (SEM).

2. Experimental

2.1 Sample collection

2.1.1 Blood collection and plasma preparation

Blood was donated by an author of this article, a 28 years old male with no record of disease. After a 12 h overnight fast, venous blood was collected at ambient temperature (24 °C) by a 21-gauge needle (length 70 mm, inner radius 0.4 mm, Microlance, Becton Dickinson, New Jersey). It was drawn into two 4.5 mL evacuated tubes (BD Vacutainers, Becton Dickinson, CA) with trisodium citrate (109 mM). The tubes were gently turned around three times and then centrifuged for 30 min, 2000 g (centrifuge Centric 400/R, Tehtnica, Železniki, Slovenia) at 22 °C in order to separate plasma from cells. The plasma supernatant was collected (down to 4–5 mm level above the cell pellet) and pooled in a 15 mL falcon tube prefilled with 5 mL of citrate-PBS buffer (154 mM NaCl, 1.4 mM phosphate, 10.9 mM trisodium citrate; pH 7.4) at ambient temperature. Isolation of EVs from this sample was performed straight after.

Due to the large required sample volume for parallel centrifugation experiments, additional blood plasma was purchased from the Blood Transfusion Center of Slovenia. The blood sample was collected according to general standards and procedures for transfusion purposes. A complete dose of blood was collected from a healthy volunteer without complications. COMPOFLOW® 4F T&B 63 mL CPD/100 SAGM–RCC–PDS-V collecting bag (Fresenius Kabi AG, Czech Republic) was used together with scale and mixer device Hemoflow 06393. The blood was centrifuged after 3 h 20 min of incubation at room temperature, centrifugation for 14 min at 22 °C at 3540 rpm (Cryofuge 6000i, rotor 6606, Thermo Scientific, ZDA) was performed (hard spin) to separate whole blood into three components: hematocrit, plasma and the buffy coat layer. The cell fractions and blood plasma were collected by CompoMat G5 and CompoMaster Net G5 System (program of separation: TB Buffy Coat). The plasma (260 mL) was not suitable for further medical utilization due to the imperfect cell elimination. The plasma was kept at room temperature (22 °C) for 20 h before dispensed into the 15 mL aliquots, frozen and stored at −80 °C.

2.1.2 EV isolation from fresh plasma

1 mL aliquots of diluted plasma were centrifuged at 17 570 g for 20 min at 22 °C (centrifuge Centric 200/R, Domel, Železniki, Slovenia, swinging rotor). 950 μL of the supernatant was saved for later analysis; pellet was washed by adding 450 μL of fresh citrate-PBS buffer and centrifuged again using the same conditions. Two EV isolate samples were fixed *in situ* (adding 50 μL of modified Karnowski's fixative to a 50 μL of sample in the tube); while the rest were resuspended in the remaining 50 μL of supernatant, and gathered (to yield app. 400 μL). Fresh buffer was added to this isolate to a total volume of 1 mL, rigorously mixed and centrifuged again using the same conditions. 700 μL of supernatant was discarded, while the pellet was suspended in the remaining liquid.

2.1.3 EV isolation from thawed plasma

To ensure the same starting plasma sample for evaluating effects of enhanced centrifugation through the experiments of parallel centrifugation (Fig. 1, Table 1: protocols #1a-b, #2-#10), plasma sample was purchased from Blood Transfusion Center of Slovenia and kept frozen at -80 °C until used. The frozen plasma was rapidly thawed at 37 °C and centrifuged at 1 000 g, 30 min, 22 °C, (centrifuge Centric 400/R, Domel, Železniki, Slovenia), to remove the remaining cells. Supernatant was transferred into another centrifuge tube and centrifuged again at 1 000 g, 15 min, 22 °C. The supernatant after second step of cell removal was then dispensed into 1 mL aliquots for isolation according to different protocols.

Table 1 List of EV isolation protocols and analyses performed on respective isolates.

Starting blood plasma sample	RCF[c]	Duration	*T*	Rotor type	Protocol	Analysis
Fresh[a]	17 500 g	20 min	22 °C	Swinging bucket	#1	FCM, SEM
Thawed[b]	17 500 g	5 min	22 °C	Swinging bucket	#1a	DLS, FCM, MS
Thawed[b]	17 500 g	30 min	22 °C	Swinging bucket	#1b	FCM, DLS
Thawed[b]	10 000 g	60 min	22 °C	Fixed angle	#2	FCM, MS, SEM
Thawed[b]	20 000 g	60 min	22 °C	Fixed angle	#3	FCM, MS, SEM
Thawed[b]	100 000 g	60 min	22 °C	Fixed angle	#4	FCM, MS, SEM
Thawed[b]	100 000 g	8 h	22 °C	Fixed angle	#5	FCM, MS, SEM
Thawed[b]	10 000 g	60 min	8 °C	Fixed angle	#6	DLS, FCM
Thawed[b]	20 000 g	60 min	8 °C	Fixed angle	#7	DLS, FCM
Thawed[b]	50 000 g	60 min	8 °C	Fixed angle	#8	DLS, FCM
Thawed[b]	100 000 g	60 min	8 °C	Fixed angle	#9	DLS, FCM
Thawed[b]	100 000 g	120 min	8 °C	Fixed angle	#10	DLS, FCM

DLS, dynamic light scattering; *FCM*, flow cytometry; *MS*, mass spectrometry. *SEM*, scanning electron microscopy.

[a]The blood plasma was prepared from fresh blood sample. Note that the donor and the plasma preparation procedure were different for this protocol than for all the rest.

[b]In order to assure the same starting sample for all tested protocols, blood plasma was purchased from Blood Transfusion Center of Slovenia. Due to the limitations of throughput of the method, the experiments were not performed on the same day. The plasma was therefore frozen and stored at -80 °C until used.

[c]Relative centrifugal forces (RCF) of the isolation and purification step of the protocol.

Isolation was performed in centrifuge Centric 200/R (Domel, Železniki, Slovenia, with swinging rotor); at 17 570 g for 5 min (protocol #1a) or 30 min (protocol #1b) at 22 °C; or in the Beckman L8-70 M ultracentrifuge (Beckman coulter, USA, with fixed angle rotor type Ti50.2) at 10 000 g (protocol #2) or 20 000 g (protocol #3) or 100 000 g (protocol #4) at 22 °C for 1 h or at 100 000 g and 22 °C for 8 h (protocol #5). 950 μL of supernatant was discarded; the pellet was resuspended in 1 mL of citrate-PBS buffer and centrifuged again. In the sample obtained by high velocity centrifugation (8 h, 100 000 g pellet), the whitish fraction just above the gelatinous pellet was resuspended in 1 mL of buffer and centrifuged again at 100 000 g for 4 h. The supernatant was discarded and the fraction above the gelatinous part (cca 50 μL) was resuspended in 450 μL of fresh buffer. The experiment was repeated at 8 °C, applying 10 000 g (protocol #6), 20 000 g (protocol #7), 50 000 g (protocol #8) and 100 000 g (protocol #9) for 1 h, and 100 000 g for 2 h (protocol #10).

2.2 Scanning electron microscopy

The samples were fixed in a modified Karnovsky's fixative (2.5% glutaraldehyde, 0.4% formaldehyde in phosphate buffer saline (PBS; 137 mM NaCl, 2.68 mM KCl, 10.14 mM Na_2HPO_4, and 1.84 mM KH_2PO_4, pH 7.4) and post-fixed with OsO_4, following the protocol adopted from Lešer et al. [47]. In brief, the supernatant after the centrifugation at 17 570 g was applied to a 0.05-μm filter (Polycarbonate (PCTE) membrane filters, ref: PCT00513100, Sterlitech, USA), lightly washed with PBS and fixed in the primary fixative for 1 h at 22 °C. The pellets were fixed in the tubes directly, or transferred to a clean cover glass and incubated for 2 h at room temperature to allow adhesion. Primary fixative (buffered glutaraldehyde and paraformaldehyde) was then added to all the samples in the ratio 1: 1, and the samples were incubated overnight at 4 °C. The fixatives were then removed by three-step washing (incubation time 10 min in each step) with phosphate buffer. Then, samples were incubated in 2% OsO_4 for 1 h, washed 3-times with distilled water (incubation time 10 min in each step), treated with saturated water solution of thiocarbohydrazide (15 min), washed with distilled water (incubation time 10 min in each step), incubated 1 h in 2% OsO_4, again; washed three-times with distilled water (incubated for10 min in each step) and dehydrated in graded series of ethanol (30%–100%, 10 min each; absolute ethanol was replaced 3 times) and hexamethyldisilazane (mixed with absolute ethanol; 30%, 50% and

100%, 10 min each), and finally air dried. The samples were Au/Pd coated (PECS Gatan 682), and examined using a JSM-6500F Field Emission Scanning Electron Microscope (JEOL Ltd., Tokyo, Japan).

2.3 Flow cytometry

FCM measurements were performed with MACS QUANT flow cytometer (Miltenyi, Bergisch-Gladbach, Germany), with the lasers set to 394 V (FSC), 454 V (SSC), and the trigger set to SSC: 1.8). Blood and platelet poor plasma were analyzed within 2 h from blood collection. The particles in fresh isolates were analyzed by flow cytometry immediately after isolation, and later, on thawed samples after up to a week of storage at 4 °C. The gates were set in a preliminary study of several blood samples; identity of cells in the presented regions was confirmed by fluorescently labeled antibodies (CD41 was used to label platelets and CD45 to label leukocytes) and by SEM analysis.

2.4 Dynamic light scattering

The dynamic light scattering experiments were performed with the 3D-DLS-SLS cross-correlation spectrometer from LS Instruments GmbH (Fribourg, Switzerland) with a 20 mV He–Ne laser (Uniphase JDL 1145P, $\lambda_0 = 632.8$ nm). The samples were equilibrated in the decalin bath at 25 °C for 15 min prior the measurement. The measurements were performed at 90° angle. All details regarding the instrumentation and data treatment are described in detail in Refs. [48–50]. The samples were diluted with the citrate-PBS in the ratio 1:10 or 1:20 to attain optimal scattering intensity and eliminate impacts of high concentration on particle diffusion. CONTIN analysis [51] was employed for transformation of correlation functions into the distribution of diffusion coefficient (D) values of species in solution. The final result of the analysis is the size distribution of particles in solution. The size parameter obtained is the hydrodynamic radius (R_h), which is calculated from the measured D by taking into account the Stokes-Einstein relation ($D = k_B T/6\pi\eta R_h$; here, k_B is the Boltzmann constant, T is the temperature and η is the viscosity of the medium). As samples were very diliute, water viscosity was considered appropriate for calculation of R_h from the measured diffusion coefficient. The R_h calculated in this way corresponds to the radius of a hypothetical hard sphere that has the same D as the particles in the sample. A representative distribution as obtained by the DLS analysis is presented in Fig. 2 for an isolate and its corresponding plasma supernatant after the first step of isolation.

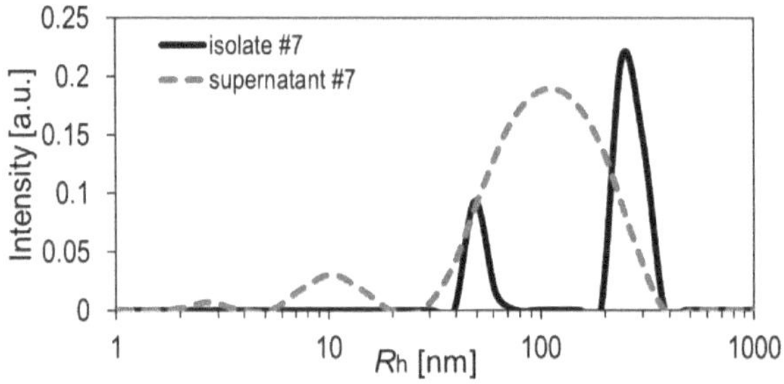

Fig. 2 A representative size distribution as obtained by the DLS analysis of an isolate and the corresponding plasma supernatant after first step of isolation. Intensity of scattering (I) given in arbitrary units (a.u.) represents the fraction of intensities I_{Rh}/I_{tot}.

2.5 Proteomic analysis

The protein contents of different isolates (obtained by centrifugation at 22 °C; assigned as isolate 1–5) were determined by Bicinchoninic acid (BCA) protein assay kit (Pierce, Thermo Scientific, USA). Volumes containing 20 μg of protein were loaded and electrophoretically separated under reducing conditions on a precast Novex Bolt 4%–12% Bis-Tris Plus gel using Bolt (3-(N-morpholino)propanesulfonic acid) MOPS SDS running buffer (Life Technologies, Carlsbad, CA, USA) according to the manufacturer's instructions, and stained with colloidal Coomassie blue. Seven bands were selected for LC-MS/MS based shotgun proteomics analysis. Details regarding MS and bioinformatics are described in Refs. [52–54].

3. Results

3.1 Low speed isolation protocol analysis

An isolate was prepared from fresh plasma sample according to the protocol #1 (Table 1) that was found to be frequently employed in clinical studies. As this protocol applies relatively low centrifugation forces (17 570 g), larger particles were expected to be present in the isolate, while the smaller ones shall be removed in the washing step.

Fig. 3 shows SEM images of the surface of the centrifuge tube in which the last step of the isolation of EVs from the fresh blood sample by the protocol #1 had been performed. Activated platelets accumulated in the bottom of the centrifuge tube (Fig. 3A), while many damaged ones were found in the region of transition from the bottom to the wall (Fig. 3B). There were many submicron particles of various shapes attached to the

Fig. 3 Scanning electron microscopy images of isolates from human blood plasma obtained by protocol #1 (17 570 g, 22 °C, 20 min, Table 1) and imaged directly in the microcentrifuge tube in which the last step of the isolation took place. (A) The platelets that were not successfully removed in the first step of isolation gather in the pellet after centrifugation at 17 570 g. The activated platelets exhibit tubular protrusions (an example exposed by the *white arrow*). (B) Damaged platelets surrounded by vesicles; the white arrow points to a corpus of a platelet with stumps that probably remained after projections were teared away during processing, (C) diversely shaped structures were found adhered to the centrifuge-tube wall. (D) High magnification micrograph of those structures. Arrows point to some characteristic shapes corresponding to the theoretical free-energy minimum conformations of membrane bilayer [17,55]: biconcave discoid (*full line arrow*), globular units joined by neck-straits (*dotted arrow*), tubes (*short-dashed arrow*), torus (*long-dashed arrow*). Scale bars in (A–C): 1 μm; and 500 nm in (D), respectively.

surface higher up in the centrifuge-tube (Fig. 3C–D). The smooth spherical endings of the observed tubular structures (Fig. 3D, short dashed arrow) suggest that these particles have membranous foundation. The pattern of their orientation ordering (see Fig. 3C) is probably caused by the flow of the sample established upon centrifugation.

Fig. 4 shows SEM images of the same EV isolate as presented in Fig. 3, but in this case, the same isolate was transferred and fixed for SEM imaging

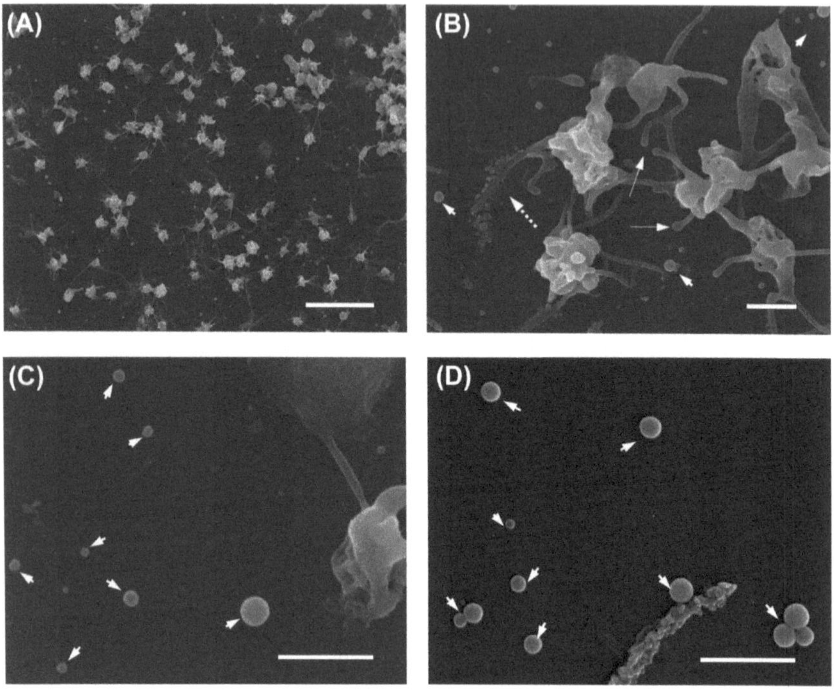

Fig. 4 Scanning electron microscopy images of isolates from human blood plasma obtained by protocol #1 (17 570 g, 22 °C, 20 min, Table 1) imaged after transferring the isolate to a cover glass. (A) residual platelets adhered to the glass surface, (B) and (C) close-ups of selected regions; (D) a coverglass (with no biological material) treated by the same fixation protocol as samples of isolates. The full-line arrows point to thickened endings of tubular protrusions. The dotted arrow shows a group of small (nanosized) particles. Arrowheads point to artifacts of fixation (a close up on (C) that were also found in blank samples (D). Scale bars: 10 μm in (A), and 1 μm in (B–D).

on a cover-glass. The activated platelets from the isolate adhered to the cover-glass (Fig. 4A). Some small structures (10–50 nm) of very regular round shape (arrowheads on the close-ups in Fig. 4B and C) were noted in the isolates. Similar structures were observed also in samples where no biological material was present (Fig. 4D), and were therefore considered artifacts of sample preparation. Only few particles with characteristic shapes of vesicles (such as the ones in Fig. 3B–D), were observed in this sample.

Fig. 5 shows the supernatant of the plasma after the first centrifugation (the part of the sample which was discarded in the process of isolation). The supernatant was transferred to the filter paper for imaging. Neither intact nor activated platelets were observed in the supernatant. Fragments of membranes (arrowheads) and some globular particles (full-line arrows)

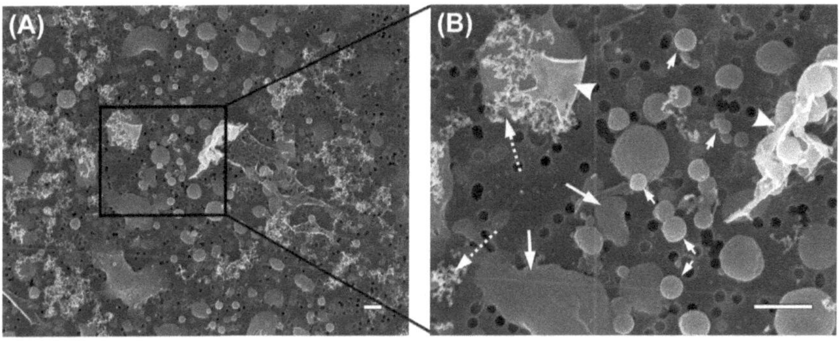

Fig. 5 Scanning electron microscopy images of the supernatant of plasma centrifuged by protocol #1 (17 570 g, 22 °C, 20 min, Table 1) – (A) and a close up of it – (B) Fragments of membranes (thick arrowheads), loose globular particles splotched on the surface of the filter (*full-line arrows*), meshwork precipitates (*dotted arrows*) and probable artifacts of fixation (*small arrows*). Scale bars: 1 μm.

were noted, many of which were splotched on the surface of the filter. Some meshwork precipitates were also observed in this sample (dotted arrows). They could have been formed during fixation by aggregation of organic components. Such structures were often observed in insufficiently washed or degraded samples (e.g. Fig. 7C and D showing the thawed sample).

Fig. 6 presents scatterplots of blood, blood plasma and isolate prepared from fresh blood sample following the protocol #1 (Table 1). Regions were determined based on preliminary tests with washed erythrocytes and samples labeled with cell-specific markers. After removal of cells (by centrifugation of blood at 2 000 g and 22 °C for 30 min), the blood plasma was rich with particles in the size range of platelets and smaller (concentration of all detected particles in cell-depleted plasma was 49 200/μL with 45% of those pertaining to the platelet region - central square gate in Fig. 6B). The scatterplot (Fig. 6C) corresponds to the isolate (concentration of detected particles was 19 620/μL with 95% pertaining to the platelet region). According to the FCM and SEM analysis, in protocol #1 (Table 1), the majority of the platelets that remained in the supernatant after centrifugation of blood at 2 000 g, were sedimented after the first centrifugation at 17 570 g, 20 min, 22 °C. It can be noted (see the scatterplots in Fig. 6) that the source blood plasma contained higher amount of particles smaller than 500 nm (the size was estimated by FCM analysis of standard polystyrene beads) than the isolate, meaning that those particles (found in supernatant Fig. 5) were largely removed in the washing steps of isolation.

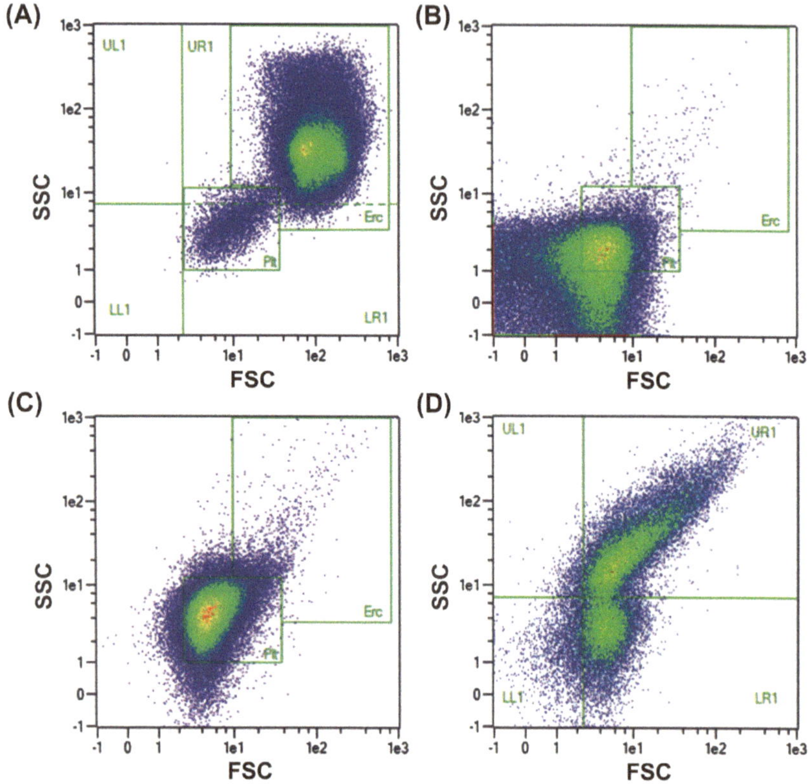

Fig. 6 Flow cytometer scatter plots of: (A) blood; (B) supernatant after centrifugation of blood at 2 000 g, 22 °C for 30 min – blood plasma; (C) isolate obtained by protocol #1 (17 570 g, at 22 °C for 20 min, Table 1); (D) thawed plasma sample, used in protocols #1a,b-#10, Table 1. Gates were set based on preliminary study (not shown). Well defined populations of erythrocytes (Erc) and platelets (Plt) can be seen (marked as Erc and Plt in Panels A–C, respectively. Leukocytes appear in the Erc gate, but in case of whole blood, their contribution to the count in this gate is negligible. Comparison of Panels B and D indicates that populations in thawed plasma sample are different from populations in fresh plasma. Populations in 4 quadrants in Panel D were defined for the convenience of the analysis: large and strongly scattering particles (UR1), large and weakly scattering particles (LR1), small and strongly scattering particles (UL1); small and weakly scattering particles (LL1). For comparison, those gates are also represented on the sample of blood (Panel A). Population detected in the plasma (B) was open-spread to small and weakly scattering events (left and below from the platelet gate, respectively). Isolate (C) appeared as a closed population with majority of particles being larger than the lower detection limit of the flow cytometer (about 400 nm).

3.2 Parallel centrifugation experiment

Fig. 7 shows SEM images of isolates from thawed plasma obtained by parallel centrifugation at different centrifugation speeds, i.e.: at 10 000 g at 8 °C (Fig. 7A and a close up on Fig. 7B, protocol #2, Table 1), at 17 570 g and 22 °C (Fig. 7C and A close up on 7D, protocol #1a, Table 1), and the surface of a pellet after 8 h centrifugation at 100 000 g, 22 °C (Fig. 7E and A close up on 7F, protocol #5, Table 1). Singular relatively intact residual erythrocytes (marked by white arrowheads) and traces of erythrocyte ghosts (marked by gray arrowheads) can be observed in the isolates obtained by lower centrifugation speed 10 000 g (Panel A). Isolates obtained by higher speeds (Panels B–D) yielded sub-micron particles (marked by dashed arrows in Panels B, C and D) and coalesced precipitates (marked by dotted arrows in Panels C and D). However, none of these particles attained a characteristic EV shape corresponding to the minimum of membrane free energy (such as ones found in isolates from fresh plasma, Fig. 3D). We could not distinguish particles in the material obtained by 8 h centrifugation at 100 000 g (Panels E and F).

The number of events in plasma and in isolates from plasma obtained by protocols #6–10 (Table 1) was assessed by FCM (the results are given in Table A1 and Fig. A3 in the Appendix). The gates for FCM analysis of thawed-plasma derived samples were set based on preliminary study and SEM analysis (Fig. 7). The events were divided among 4 quadrants (shown in previous subchapter, Fig. 6D and in Appendix, Figs. A1–A2, A5); upper-right (UR): residual erythrocytes, cell ghosts (wide arrowheads on Fig. 7), and large aggregates; lower-right (LR): fragmented cells (full line arrows in Fig. 7); lower-left (LL): small and weaker side-scattering particles; upper-left (UL): small but stronger side-scattering particles). For analysis convenience, events were divided between larger particles (right regions: cell residual) and smaller particles (left regions).

Fig. 8 shows FCM records on the number of total events (A) and in of events recorded in the gates where we expect residual cells and cell debris (B), and smaller particles (C). Results on parallel centrifugation at increasing centrifugation speeds are shown. The number of events in the gates where we expect residual cells fluctuated for about 25% in source plasma (black bars in Fig. 8B) which estimates of the FCM measurements.

The number of events was higher in isolates than in other samples. The number of events was higher in plasma than in supernatants, but the clearance of particles was not directly related to applied centrifugation speed. No

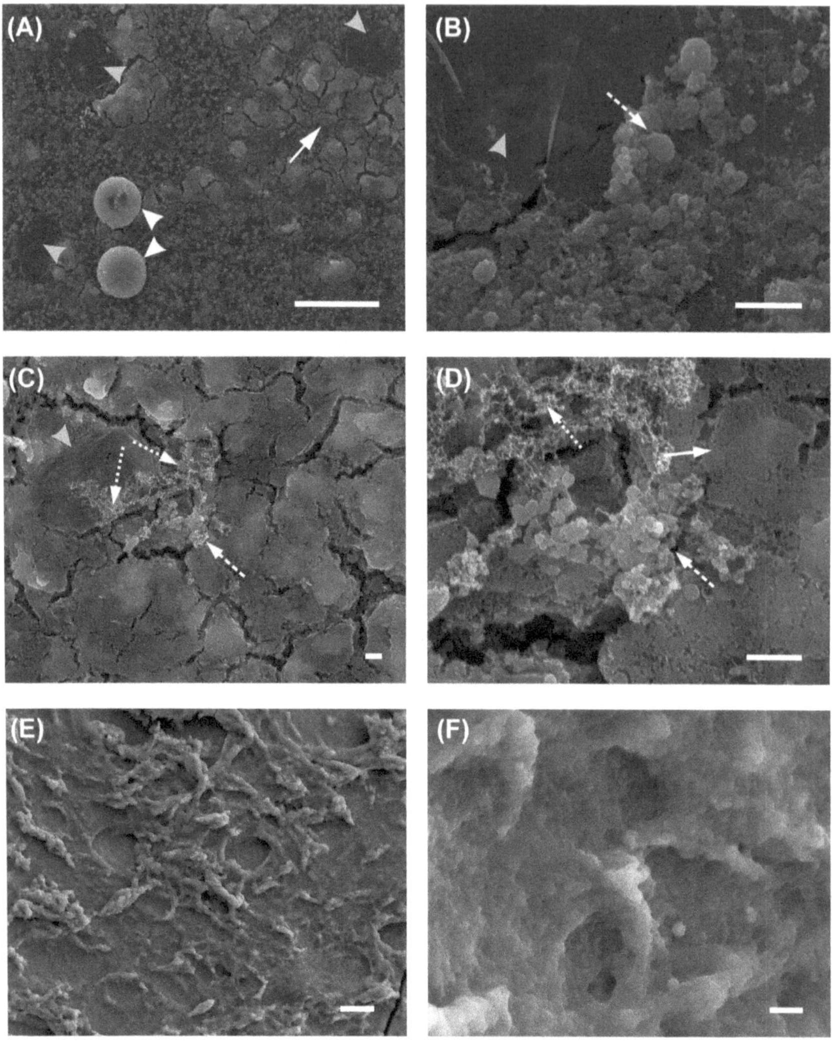

Fig. 7 Scanning electron microscopy images of isolates obtained by centrifugation at 10 000 g at 8 °C (A and a close up in B, protocol #2, Table 1), at 17 570 g and 22 °C (C and a close up in D, protocol #1a, Table 1), and the surface of a pellet after 8 h centrifugation at 100 000 g (E, and a close up in F (protocol #5, Table 1). The wide arrowheads point to erythrocytes (white arrowheads) and erythrocyte ghosts (emptied cell membranes, gray arrowheads). Full-line arrows indicate platelet remains (degraded deposits of aggregated cells), dashed arrows point to sub-micron-sized particles (EVs), while dotted arrows in Panels B–D point to nets of small globular particles, presumably precipitates of proteins. Material in pellet obtained after 8-h centrifugation at 100 000 g was of inrecognizable structure (pannels E, F). Scale bars: 10 μm in (A), and (E), 1 μm in (B), (C), (D) and (F).

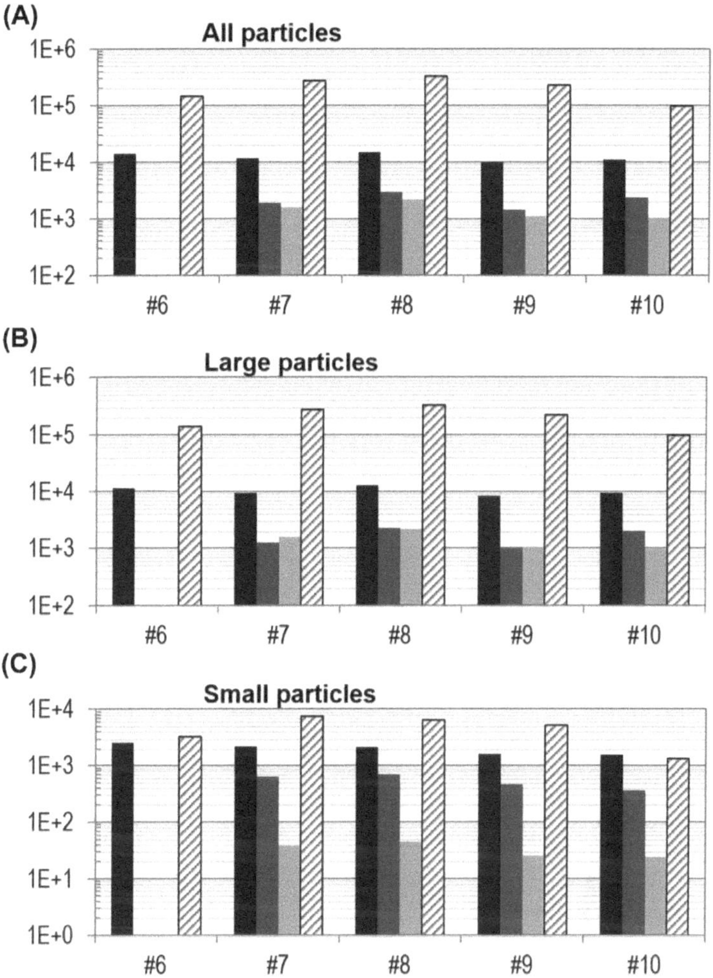

Fig. 8 Concentration of particles in samples from plasma, isolation supernatants and isolates as measured by flow cytometry. We compared protocols #6–10 (Table 1) presenting increasing speed of parallel centrifugation (#6: 10 000 g, 1 h, #7: 20 000 g, 1 h, #8: 50 000 g, 1 h, #9: 100 000 g, 1 h, #10: 100 000 g, 2 h). Events detected by FCM (A) were divided with respect to their FCS signal, which is proportional for the particle size. Changes in larger particles' concentrations are represented in graph B and those in smaller particles in Panel C. Supernatants of the sample processed by protocol #6 have not been measured due to technical problems. Black – source plasma samples; dark gray – first supernatant; light gray – second supernatant (washing step); stripped bars – isolate.

trend was observed in the number of events corresponding to platelets and larger fragments in the supernatants with increasing centrifugal pull. The number of events corresponding to larger particles was higher than the number of events corresponding to smaller particles in all samples (see that the scales in panels A and B differ by 2 orders of magnitude). Interestingly, the number of larger particles was almost the same in the two supernatants (Fig. 8B) while the number of smaller particles (EVs) was smaller in the second supernatant (Fig. 8C). This indicates that larger fragments were pelleted in both centrifugation steps while some smaller ones were lost in the first step of isolation (plasma centrifugation).

The respective number of events corresponding to both, larger and smaller particles in isolates, increased with enhanced centrifugal pull and reached a maximum at 50 000 g (Fig. 8A–C). With continuing increase of the pull, the number of events then decreased at 100 000 g, 1 h (protocol #9) and further dropped at extended duration of UC at 100 000 g, 2 h (protocol #10).

In the first and second supernatant there was an interesting inversion of the ratio between weaker and stronger side-scattering particles (Appendix, Fig. A4). In the first supernatant there was a larger share of weaker side-scattering particles, while in the second supernatant, strongly side-scattering particles represented a vast majority in all samples. It seems that some particles are eliminated with the supernatant based on their buoyancy (density and shape).

As by the FCM small and large particle concentration appeared proportional to each other, the possible hidden differences in smaller particles were assessed by DLS analysis.

The results of DLS analysis of the isolates and their corresponding plasma supernatants after the first step of isolation (high speed centrifugation) are given in Appendix (Table A2). Note, that the size parameters obtained by DLS should not be taken as direct dimensions of the particles, rather they must be considered with respect to particle's scattering properties. Furthermore, the width of a peak (dispersion of particle sizes around the determined mean size) should be taken into account as well.

The size distribution of particles in the isolates in comparison to plasma sample and supernatants showed similar trend in all samples. In Fig. 2 (in the Experimental section), prevalence of smaller particles ($R_h < 50$ nm) in plasma supernatant over those in the isolate can be observed. This is expected, as free plasma proteins ($R_h < 10$ nm) are presumed to remain in the supernatant, and the isolate is supposed to be further cleared of them in the washing step(s). The peaks with mean R_h 10–500 nm were in focus in this study, as they are presumed to carry information about vesicles (peaks with R_h 10–200 nm) and cell debris (peaks with R_h 200–400 nm, particles assessed by FCM, Fig. 8). The intensity of scattered light was determined for each separate peak. Intensities of the peaks with $R_h \leq 1000$ nm were counted-up and their contribution to the total measured intensity was calculated (the results are given in Table A2). The smaller contribution of particles with $R_h > 1000$ nm to the total scattered light intensity in the isolates prepared with higher centrifugation speeds therefore indicates the higher presence of large aggregates ($R_h < 5$ μm) in them.

Fig. 9 shows the intensity of scattered light measured by DLS analysis and concentration of particles assessed by FCM. As the trends were similar for the peaks of smaller particles (with mean R_h (10–100) nm, and for the larger particles with mean R_h (100–500) nm; the collective intensity of scattered light for all the particles with $R_h \leq 1000$ nm is therefore shown in Fig. 9 ($I_{\mathrm{Rh} \leq 1000} = \int_{Rh=0}^{Rh=1000} I(Rh)dRh$, gray bars). Stronger scattering was detected in samples with higher concentration of particles determined by FCM (see the correspondence of DLS and FCM results presented in Fig. 9 by gray and black bars). Because the large particles scatter light much stronger than the small ones, and because the particles detected by FCM are included in the populations assessed by DLS, this agreement of

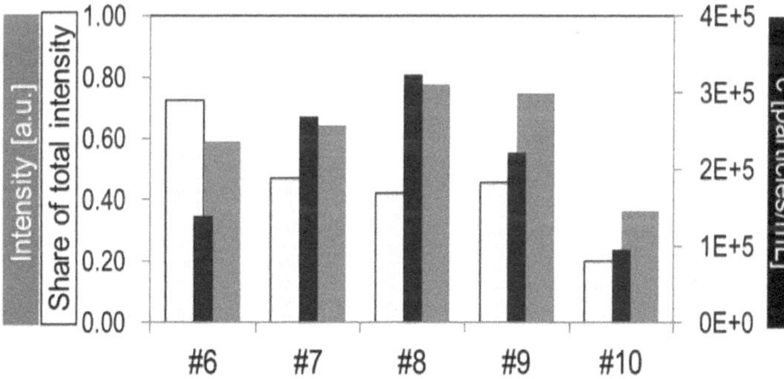

Fig. 9 Comparison of dynamic light scattering (white gray and gray bars) and flow cytometry results (black bars) for isolates obtained by protocols #6–10 (Table 1, increasing centrifuge speeds from 10 000–100 000 g at 8 °C). The intensity of scattered light was determined for each separate peak (I_{Rh}). Intensity of peaks with $R_h \leq 1000$ nm was summed-up ($I_{Rh \leq 1000} = \int_{Rh=0}^{Rh=1000} I(Rh)dRh$, gray bars) and their relative contribution to the total measured intensity was calculated (fraction of intensities $I_{Rh \leq 1000}/I_{total}$, white bars). Concentration of all detected particles by FCM is demonstrated by black bars. a.u. – arbitrary units; c – concentration; R_h – hydrodynamic radius.

results is expected. On the other hand, this correlation indicates that more severely processed samples did not contain larger amounts of small particles, which are below the detection limit of FCM, but can be detected by DLS. The decrease in the contribution to the total intensity of scattered light attributed to the particles with $R_h \leq 1000$ nm (fraction of intensities $I_{Rh<1000}/I_{total}$, presented in Fig. 9 by white bars) rather suggests that increasing forces at higher centrifugation speeds cause particle aggregation. Larger particles scatter light much stronger than the small ones, and therefore, even very few aggregates (particles $R_h > 5$ μm) that are present in the sample strongly lower the contribution of smaller particles ($R_h \leq 1000$ nm) to the total intensity. In contrast, very large particles may be excluded from FCM detection as they exceed the detection range determined by the settings, or if these particles are associates of many, they might dissociate in the flow before the detection.

FCM displayed a trend in the change of side scattering properties of particles with increasing centrifugation pull. While there was little or no change detected in the forward scattering signal (Fig. 10A–E, K-O), the side scattering of the particles in the isolate became stronger with the increase in the centrifugation pull (see the SSC scatterplots in Fig. 10F–J, P-T). The single peak with relatively low SSC gradually broadened toward right (Fig. 10F–J) and split at higher forces (protocol #10, centrifugation at 100 000 g, 2 h, 8 °C, Fig. 10P). This trend was further amplified when centrifugation was performed at higher temperature. The peak with higher SSC enhanced and prevailed in protocols #2–4 (1 h centrifugation at 22 °C at 10 000 g, 20 000 g, and 100 000 g, respectively; Fig. 10Q–S), and finally disappeared in the isolate of protocol #5 (8 h centrifugation at 100 000 g, 22 °C; Fig. 10T). In the sample processed for 8 h, the drop in the scattering intensity (shift of FSC and SSC distributions to the left, Fig. 10O and T) suggests degradation of the sample.

3.2.1 *Analysis of the changes in protein profiles*

The concentrations of proteins in the isolates obtained by the protocols #1a-5 (from the same blood plasma sample) determined by the BCA method were 2.30 (isolate #1a), 4.01 (isolate #2), 17.22 (isolate #3), 15.55 (isolate #4) and 101.08 mg/mL (isolate #5). The SDS-PAGE image of their protein profiles (20 μg proteins) is shown in Fig. 11.

The overall protein profiles of the various isolates are very similar and characterized only with slight expression differences. Some well-resolved bands were selected for MS-based proteomic analysis as shown in Fig. 11.

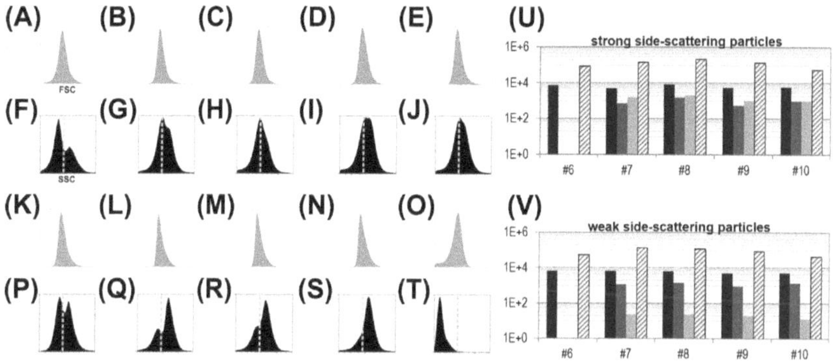

Fig. 10 Forward (gray) and side scattering distributions (black) of particles in the source plasma sample and in the isolates obtained by centrifugation at different centrifugation speeds at 8 °C (protocols #6–10, Table 1) and at 22 °C (protocols #2–5, Table 1) as determined by FCM. Please note, that in protocols #2 and #6; #3 and #7; and #4 and #9, same speeds but different temperatures were applied. (A, F) a plasma sample; (B, G) isolate of protocol #6, centrifugation at 10 000 g, 1 h, at 8 °C; (C, H) isolate of protocol #7, centrifugation at 20 000 g, 1 h, at 8 °C; (D, I) isolate of protocol #8, centrifugation at 50 000 g, 1 h, at 8 °C; (E, J) isolate of protocol #9, centrifugation at 100 000 g, 1 h, 8 °C; (K, P) isolate of protocol #10, centrifugation at 100 000 g, 2 h, 8 °C; (L, Q) isolate of protocol #2, centrifugation at 10.000*g*, 1 h, 22 °C; (M, R)isolate of protocol #3, centrifugation at 20 000 g, 1 h, 22 °C; (N, S) isolate of protocol #4, centrifugation at 100 000 g, 1 h, 22 °C; (O, T) isolate of protocol #5, centrifugation at 100 000 g, 8 h, 22 °C. The plots are normalized to surface area = 1. A vertical guideline is added to the SSC plots F-J and P-T to highlight the trend of change in side-scattering properties of particles in the isolates obtained by different protocols in the presented order. Changes in concentration of strongly (U) and weakly (V) scattering particles throughout the processing by different protocols are also represented (supernatants of the sample processed by protocol #6 have not been measured due to technical problems). Black bars – source plasma samples; dark gray bars – first supernatant; light gray bars – second supernatant (washing step); stripped bars – isolate.

The results of LC-MS/MS-based protein identification are shown in Table 2. Proteins identified in the various samples are highly abundant human blood plasma proteins, as explained by Strohkamp et al. [63]. Proteins identified in bands 1–3, for example, were found to be progressively enriched in samples #2, #3 and #4 upon increasing centrifugal forces. By comparing the isolates prepared with different centrifugation speeds, we see accumulation of fibronectin (Band 2) with increasing force. Fibronectin was largely eliminated in the isolate #5. Protein(s) in Band 5 were highly enriched in the sample #4 (8 h centrifugation time) while Apolipoprotein-B-100 (Band 6) decreased with increased centrifugation speed. Other bands did not differ substantially in the different samples.

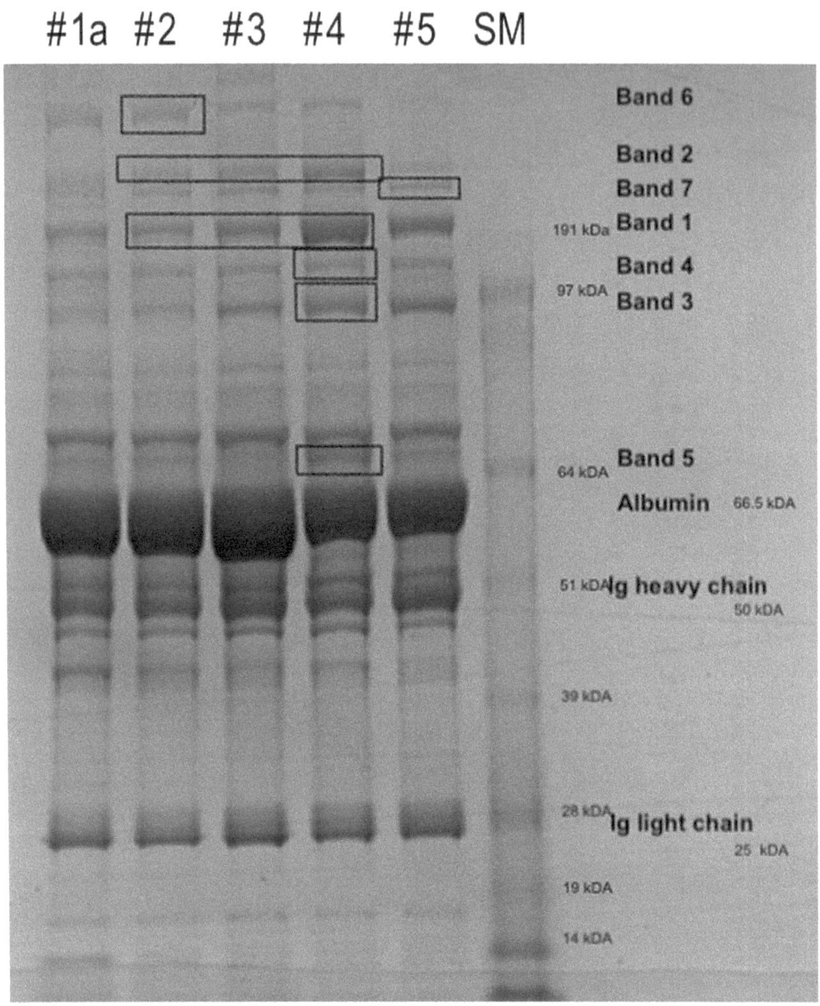

Fig. 11 SDS-PAGE protein profiles of isolates obtained by protocols #1a-5 (different centrifugation times and speeds, performed at 22 °C) at reducing conditions. SM – size marker. Band selection for mass spectrometry analysis is noted by the frames. Bands 1 and 2 were taken from the isolate #4.

4. Discussion

4.1 Source of vesicles in the isolates

From a very basic perspective, two springs of EVs can be expected in biological samples – i) internal – cell-directed production of vesicles through controlled change of membrane components (including formation

Table 2 Gel-based proteomics results. Proteins identified in isolates 1a-5 using different centrifugation regimes (protocols #1a-5). UniProt ID indicates the accession numbers from the UniProt database; Mascot ion scores indicate the scores obtained from the Mascot software. Identification of protein in previous proteomic research of EV isolates [64].

Band	UniProtID	Protein name	Molecular weight (kDa)	Score	Previously related to EVs
1	P01023	Alpha-2-macroglobulin	165	1287	[56–59]
2	P02751	Fibronectin	266	498	[56,59–61]
3	P01023	Alpha-2-macroglobulin	165	509	[56–59]
	P01024	Complement C3	189	302	[56,58,59]
	P02768	Serum albumin	71	80	[59]
4	P00450	Ceruloplasmin	123	121	[56,60,61]
	P01859	Immunoglobulin heavy constant gamma 2	37	71	[62]
5	P01871	Immunoglobulin heavy constant mu	50	443	[56,60,61]
	P01024	Complement C3	189	346	[56,58,59]
	P02768	Serum albumin	71	324	[56–61]
	P0C0L4	Complement C4-A	194	276	[56,59]
	P04003	C4b-binding protein alpha chain	68	120	[59]
6	P04114	Apolipoprotein B-100	517	495	[56]
7	P02751	Fibronectin	266	79	[56,59–61]
	P02768	Serum albumin	71	77	[56–61]
	P01023	Alpha-2-macroglobulin	165	69	[56–58]
	P02760	Protein AMBP	40	58	[56,59,60]

of so-called ectosomes, exosomes and apoptotic bodies); and ii) external — spontaneous vesiculation caused by membrane perturbations under the change of system conditions (mechanical stress, change of chemical environment, temperature, etc). Considering the external path of EV genesis, the amount of EVs in an isolate depends on the quantity of cells and their fragility; while in case of internal path, the amount of EVs is expected to reflect metabolic (including communication) activity of the cells in focus. The ratio between vesicles of a certain cell origins is presumed to be related to the abundance of the cell populations in the examined tissues/samples and their vesiculation-activity.

In case of blood, blood cells are subjected to various perturbing conditions from the sampling on, e.g. change in temperature, contact with exogenous surfaces, shear forces in the needle etc. All of those may contribute to cell fragmentation and activation. It was previously observed that large portion of vesicles isolated from blood is of the platelet and/or erythrocyte origin [65—70]. Erythrocytes are by far the most aboundant cells in blood, high prevalence of vesicles of their origin is therefore expected in case of cell degradation. Platelets are second most aboundant cells in blood. Further, they are prone to vesiculate, especially in the activated state [68,70,71]. If a sample contains cells [72], vesicle count depends on sample storage. The studies showed, that the amount of platelet-derived EVs greatly increased when samples with platelets were exposed to lower temperatures, which can be related to platelet activation [17,68,73].

In the isolation processing the particles move, and are exposed to shear stress, which also provokes vesiculation [17,43]. One very important factor in centrifuge processing is temperature. Besides possible cell activation, lower temperature increases the viscosity of the sample — altering sedimentation times of particles, increasing shear forces, and changing membrane mechanical properties [74,75].

Some cells remain in the supernatant after the first centrifugation step, and most of them are platelets (represented in the scatterplot in Fig. 6B, and Figs. 3 and 4). Platelets are the smallest blood cells, and upon activation, they get branched (see Figs. 3A or 4A), which contributes to their slower sedimentation rate. In the following steps of isolation, the retained cells are subjected to high-speed centrifugation (i.e. > 10 000 g). In our samples, majority of the residual platelets sediment at 17 570 g. Interestingly, the platelets in the pellet retained adhesive properties as they bound to glass in the process of preparation for imaging (Figs. 3A and 4A). The morphology of adherent platelets in the transferred samples (activated globular shapes with tubular protrusions; Fig. 4A) may be induced by the action of shear forces during the processing, although similar morphology was reported in relation to their interaction with glass surface [76] or their late apoptotic state [77].

When cells and their fragments during centrifugation reach the tube wall the shear forces on them greatly increase. If the cells are fragile and their membranes are fluid enough to form vesicles after tearing, the vesicles are formed from damaged cells. This can result in high quantity of vesicles originating from cells that were not efficiently removed in the previous steps of isolation.

In the supernatant after centrifugation of fresh plasma at 17 570 g (protocol #1), fragments of membranes (arrowheads, Fig. 5) and some loose globular particles (full-line arrows, Fig. 5) splotching on the surface of the filter were noted. We speculate that they could be low density lipoproteins, lipid droplets or possibly some low density vesicles, formed after cell destruction and membrane phase separation. Such vesicles are not expected to sediment due to possible low density content, poor on proteins. However, those particles were not exhibiting predicted vesicle-shapes.

Microparticle formation upon shear stress was reported by many authors, e.g. Ref. [17,43,78–80]. Similarly to the previously published data [17,43], we found many vesicles attached to the surface of the centrifuge tube wall. Their shapes (see Panels B–D of Fig. 3) correspond to characteristic shapes calculated by the minimization of the membrane free energy [17,55], which indicates that they are without internal structure and fulfill the requirements for the definition of cellular vesicles. Tubular particles seemed to orientate according to the flow established during centrifugation (see Fig. 3C), and were likely formed due to shear forces. Similar platelet tethers related to shear stress were described by Reininger et al. [79]. Round vesicles can be formed later from such tubular remnants by "pearling" [79,81]. Indication of such proceeding is visible in Fig. 4 (see short-dashed and dotted arrow). As vesicle-like particles were found in high amounts only near the tube surface, we hypothesize that they were formed from the platelets in the high shear stress field near the centrifuge-tube wall. Large share of vesicles of platelet origin in isolates can therefore be ascribed to platelets' proneness to form vesicles, and their lag behind in sedimentation, bound up to their later fragmentation in the subsequent processing. We assume that the observed structures found in the isolates by protocol #1 are representative also for other protocols considered (if fresh blood is used as a source material), meaning that such particles, formed upon processing may overwhelm the isolates, claimed to be extracellular vesicles.

The isolation protocols yield material, which upon analysis (e.g. by light scattering and microscopic techniques) reveals presence of particles. The samples prepared from fresh and frozen/thawed blood plasma were analyzed in this study with the aim to better understand the origin of these particles found in the isolates. Thorough analysis of individual steps of the isolation protocols was performed. As EVs are too small to be observed by optical microscopy, the samples were imaged by scanning electron microscopy. Dynamic light scattering was employed to reveal the size distribution of particles, and flow cytometry was used to determine concentrations of particles larger than about 400 nm. In analysis of the isolate prepared from fresh plasma by a protocol commonly used in clinical studies [17,32–46] many activated platelets were observed in the sample mounted on holder, while the deposits of vesicles with characteristic shapes of minimal membrane free energy were found attached to the walls of the centrifuge tube in which the isolation was performed (Figs. 3 and 4). Present and previous studies indicate that these vesicles are formed by tearing and re-sealing of the residual cells in the samples (mostly erythrocytes and platelets). The present study provided further evidence that the isolates of a fresh plasma by the protocol that includes centrifuge centripetal accelerations up to about 20 000 g consist of a heterogeneous mixture of residual platelets and blood cell fragments.

Frozen/thawed samples were treated at different centrifugation speeds, including the one similar to the one used for the fresh sample (17 570 g/ 20 000 g), but also milder and harsher ones, in order to reach the renown conditions for gathering the exosomes at around 100 000 g. It was found that all the isolates from frozen/thawed samples were substantially different from the isolates from fresh samples (compare Figs. 3, 4, and 7). The heterogeneous population of submicron-sized particles from frozen/thawed plasma observed in micrographs (Fig. 7) could not be recognized as a population of intact or even activated platelets; the shapes of observed particles were of irregular globules. Aggregates of nanoparticles were presumed to be protein precipitates (Fig. 7). No particles with characteristic shapes corresponding to the membrane free energy minimum that would comply to the definition of cellular membrane vesicles (membrane enclosed entities without internal structure, such as the ones evident in Fig. 3D in fresh plasma) were found in the samples from thawed plasma. We suggest that the majority of particles found in the isolates of frozen/thawed plasma are residual platelets that deformed and/or fragmented due to thermal and mechanical stress imposed upon them in the process of isolation. Both techniques, FCM and DLS detected particles in the isolates. There was an agreement of the estimated concentration of particles trend detected by these two techniques (Fig. 9) showing that the concentration of particles in the isolates varied with changing the parameters of processing (the centrifuge centripetal acceleration). These results provide further evidence in favor of the hypothesis that majority of the isolated particles are fragments of residual cells in samples. Most importantly, DLS measurements have shown differences in size distribution and amount of particles with hydrodynamic radii smaller than 1 μm with different centrifugation pull (Appendix Table A2) that were not correlated to the clearance of same sized particles from the sample supernatant. This indicates that the small particles in the range of expected "exosomes" may be created in the process of isolation.

We failed to observe EVs in the exosome size range in the SEM images of the thawed samples. However, we also did not succeed to observe larger EVs when the isolate was transferred for imaging on a filter paper or on glass (Fig. 4) and there is a possibility that smaller particles were lost during mounting of the sample on the stands for SEM. Further improvement of sample preparation is necessary to give a decisive answer on the question whether the EVs with characteristic shapes of membrane-enclosed entities smaller than 100 nm are present in the isolates from human blood plasma with the protocols for harvesting exosomes.

These observations provide important knowledge as regards DC protocols and size and shape of vesicles. Commonly, the strategy for larger vesicles harvesting is application of lower centrifugation speed, while higher speeds (and forces) are applied for isolation of smaller vesicles [82]. This seems rational according to the sedimentation theory, but considering shear-induced tearing, it is also expected that stronger forces, will cause stronger grinding and that the size of particles (obtained from the same starting material) will therefore depend on centrifugation speed and time. The FCM analysis in this study shows that the material is being transformed throughout the centrifugation processing. One can observe the change of populations which correlate with constraining of the applied conditions (first increase of speed and later increase in temperature). While hardly any change can be observed in FSC, the differences in SSC properties of the particles in the samples are apparent. We could not attribute this distribution shift to the additional (smaller and lighter) particles being pelleted in samples of enhanced UC procedure. It therefore seems that it is related to the transformation of material upon processing. The stronger scattering particles appeared to be more stable, and were even further produced along the sample storage (see Appendix; Fig. A5).

4.2 Isolation efficiency

Understanding of sedimentation is crucial for efficient isolation of vesicles by DC. The properties of vesicles and the medium must be considered in the protocol design. The sedimentation rate of particles depends on their density, size and shape, and the viscosity of medium they are passing. During the centrifugation, components of the sample arrange/separate due to centrifugal forces and establish a density gradient. Particles travel until the equilibrium state is established (sedimentation limit is reached when the particles are distributed by their density).

Although the increase of speed and time of centrifugation reinforce the efficiency of sedimentation, long time/high speed centrifugation may cause vesicle degradation [12,25,26]. Our experiments suggest that redistribution of lipids and proteins may occur in prolonged UC. The particles in our samples lost their optical density after centrifugation at 100 000 g for 8 h (see Fig. 10T). The nucleic acids and proteins may severely precipitate at such extreme conditions and form a gelatinous precipitate.

Temperature is a very important factor in EV isolation processing. As long as the cells are present in the sample, the change in temperature may induce their activation. On the other hand, it also strongly affects medium viscosity, and mechanical properties of sample components. Medium viscosity is lower at higher temperature; the sedimentation is faster and its limit is reached sooner. While lower temperatures are expected to preserve proteins and make membranes more rigid, there are indications that loss of cell membrane fluidity, makes it more susceptible to mechanical damage [83]. Furthermore, shear forces which are key in fragmentation of cells in the centrifuge are higher at lower temperatures.

Cell fragmentation is considered as an artifact in isolation protocols. To avoid cell fragmentation, it is recommended to remove the cells and large particles in several low-speed steps [15]. However, when the supernatant is transferred, some vesicles from the sample are lost due to their adhesion to the surfaces of the centrifuge tube and pipette tips. These losses depend on the composition and properties of the membranes as well as on the properties of the material composing laboratory equipment. Therefore, the procedure and materials used may also modify the populations of vesicles in the final isolate. In our study, microvesicles were found to stick quite strongly to the polypropylene surface, and were largely missing in the transferred sample (represented on Fig. 3 of the isolate in the tube, and Fig. 4 of the isolate fixed on cover glass).

4.3 Identity of particles in EV isolates (EVs' genuineness)

Finding specific vesicle markers is challenging because, regardless of them being intact or re-formed, the vesicle membrane retain the membrane associated proteins. It is an essential task to evolve purification to ensure that free proteins, nucleic acids etc. acquired during the processing do not shade those which are innately integrated-in or exposed-on EVs.

From Fig. 11, it is clearly apparent that the protein profiles of all our samples were very similar to that of starting whole plasma sample. The protein concentration of isolates was consistent with the concentration of the particles determined by FCM, while DLS has shown that small particles (with $R_h < 15$ nm) were much more abundant in the supernatant fractions than in the isolates. These observations suggest that a large portion of proteins in these isolates is a part of supramolecular assemblies, rather than dissolved as single particles alone.

The amount of apolipoprotein B-100 (APOB, band 6) was found to be decreased in the isolates prepared with increased time of ultracentrifugation. Accounting APOB inclusion in low density lipoproteins (chylomicrons, VLDL and LDL), it is expected to accumulate in the top layers of the supernatant. The efficiency of APOB removal is therefore correlated with the establishment of the density gradient in the sample upon UC.

Alpha-2-macroglobulin (A2MG, bands 1 and 3) was found to be enriched in pellet fraction proportionally to the keener conditions. A2MG is known to display different SDS-PAGE bands under reduced conditions; moreover, it is sensitive to freezing and can be degraded during sample processing. The increased intensity of band 1 could be due to UC related efficiency of sedimentation which is in-turn associated with enhanced protein degradation.

Fibronectin (FINC, band 2) was enriched in the isolates of protocols #1a-4. Activation of hemostasis system can induce fibronectin binding to platelets [84,85] and may therefore sediment with membrane particles. On the other hand it is known that shear stress can induce its unfolding and polymerization [86,87]. It is therefore probable, that in the isolate #5, the fibronectin was a part of gelatinous pellet and largely excluded from the isolate.

Band 5 was disproportionally augmented in the isolate of protocol #4. It was found to contain several proteins that are associated with immune system activation, namely labeling of damaged material for clearance. Other bands did not differ substantially in different samples.

All the identified proteins of the isolates obtained in protocols #2—4 were previously associated with EVs, blood derived microparticles, or cell membranes (the associations denoted in Ref. [64] are summarized in Table A3). It is generally known that various "contaminants" such as non-EV proteins are found in EV isolates, and it was also seen that EVs isolated by UC are associated with a substantial amounts of albumin and other plasma proteins [56—61,88].

There are various possible causes for specific plasma proteins to be enriched in the pellet fraction (isolate) besides being a genuine constituent of EVs. For example, some proteins or protein aggregates can (a) sediment due to their size, shape and density, (b) form complexes with higher sedimentation rate, (c) bind to cell membranes and sediment together with them; (d) get enclosed into the cell fragments, formed upon cell mincing during processing. Whether the presence of those proteins in EV isolates is a feature of impurity of the sample or innate is yet to be clarified. The letter would mean that we may need to change our conception of "free" (soluble) plasma proteins.

We suppose that certain protein enrichment in isolates can be a result of up-taking surrounding components after reformation of the disrupted membranes. It was previously reported that plasma microparticles are hardly cleaned of [21] or bind [89] fibrinogen; which might be related to shear stress- induced vesicle formation and fibrinogen adhesive conformation change [90]. EVs are also frequently reported to expose phosphatidylserine (PS) [15,91] and bind complement system [92], as was also found in our samples. This is intuitively contradictory for particles striving long-distance reach, as such particles are expected to be quickly discarded by the macrophages [72,93]. Yet, PS is one of the generally accepted EV markers [15,91], and little was considered about such particles being a side product of isolation. Arraud et al. [66] on the other hand showed that not all vesicles in blood plasma expose PS, and suggested that vesicles exposing PS are either formed by cell rupture or degradation, and subjected to the loss of membrane asymmetry.

Our results show that vesicles formed in fresh plasma during processing have characteristic shapes which are usually non-spherical (see Fig. 3). Also, there seems to be a certain type of optically dense particles (strong side scattering) that can be produced during centrifugation (see Fig. 10) and sample storage (Appendix, Fig. A5), that are far more stable than others, and are therefore expected to be found in majority of the non-fresh blood derived EVs samples.

5. Conclusions

Better understanding of EVs' physicochemical properties is needed to further reveal their biology and exploitation of their potentials. In light of the hardly awaited break-through of extracellular vesicles from "could be used as" to "are successfully used in/as/for", we here discoursed some factors hindering the progress in consideration of centrifuge-based isolation protocols. It seems that reporting observations is no longer enough, and further progress demands that we weight up what those observations mean. The close inspection of centrifugation processing in this study implies that starting material is easily transformed. While temperature, shear, and pressure play important parts in processing, uniqueness of starting samples defines the resulting isolate. This work 94 provided evidence that the majority of particles found in isolates from blood plasma are fragmented blood cells, mostly platelets whereby fragmentation takes place due to thermal and mechanical stress during processing. Freezing/thawing of samples made essential changes to the isolate. Efficiency of isolation was found to reach a plateau between centrifuge accelerations between 50 000 g and 100 000 g. Prolonged centrifugation at 100 000 g was found to cause aggregation and degradation of material. Further and more detailed analysis is in progress [94].

6. Appendix

6.1 Flow cytometry

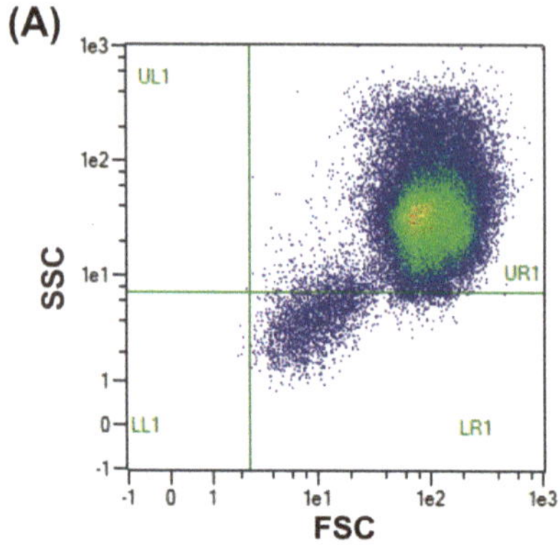

Fig. A1 Examples of events corresponding to cells and particles in fresh blood (A) and in the platelet-rich plasma (B) prepared from it. Events corresponding to the erythrocytes, leukocytes and cell debris are found in the upper-right region (UR1) of the FSC-SSC scatterplot, while majority of platelets appear in the lower-right (LR1) region. The left two quadrants (UL1 and LL1) represent smaller and weaker-scattering particles, respectively.

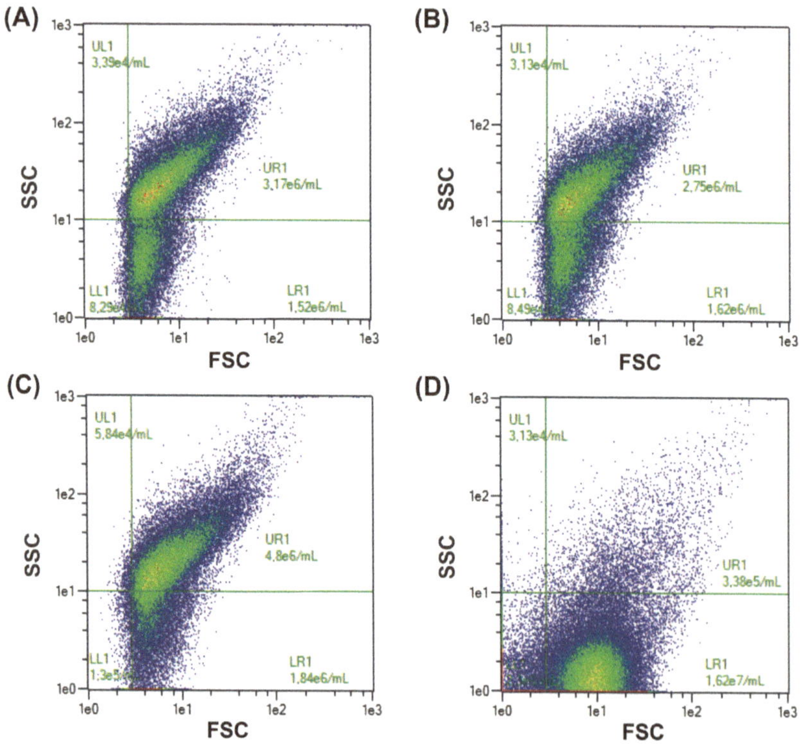

Fig. A2 Examples of FSC-SSC scatterplots pertaining to the EV isolates from a frozen/thawed plasma sample, centrifuged at 22 °C: (A) 10 000 g, 22 °C for 1 h (isolate #2); (B) 20 000 g at 22 °C for 1 h (isolate #3); (C) 100 000 g at 22 °C for 1 h (isolate #4) and (D) 100 000 g at 22 °C for 8 h (isolate #5).

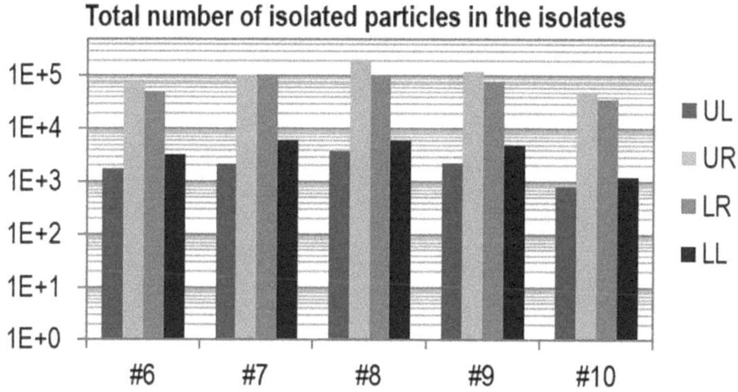

Fig. A3 Comparison of particles in different gates in isolates #6–10. *UR*, large and strongly scattering particles; UR – large and weakly scattering particles; UL – small and strongly scattering particles; LL – small and weakly scattering particles.

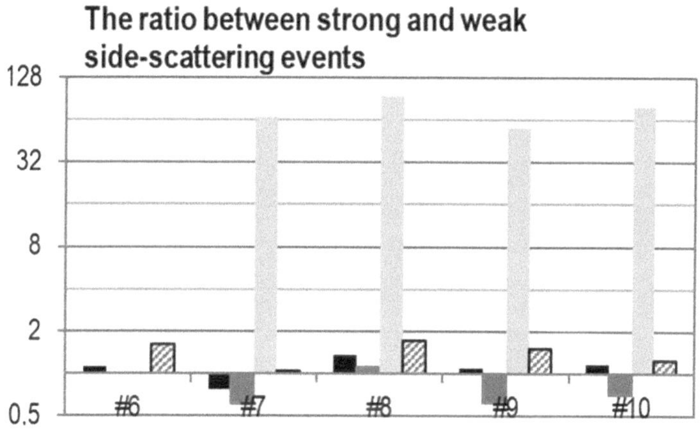

Fig. A4 Comparison of the ratios of strongly and weakly scattering particles in the samples obtained in the processing according to isolation protocols #6–10. Black bars – source plasma samples; dark gray bars – first supernatant; light gray bars – second supernatant (washing step); stripped bars – isolate.

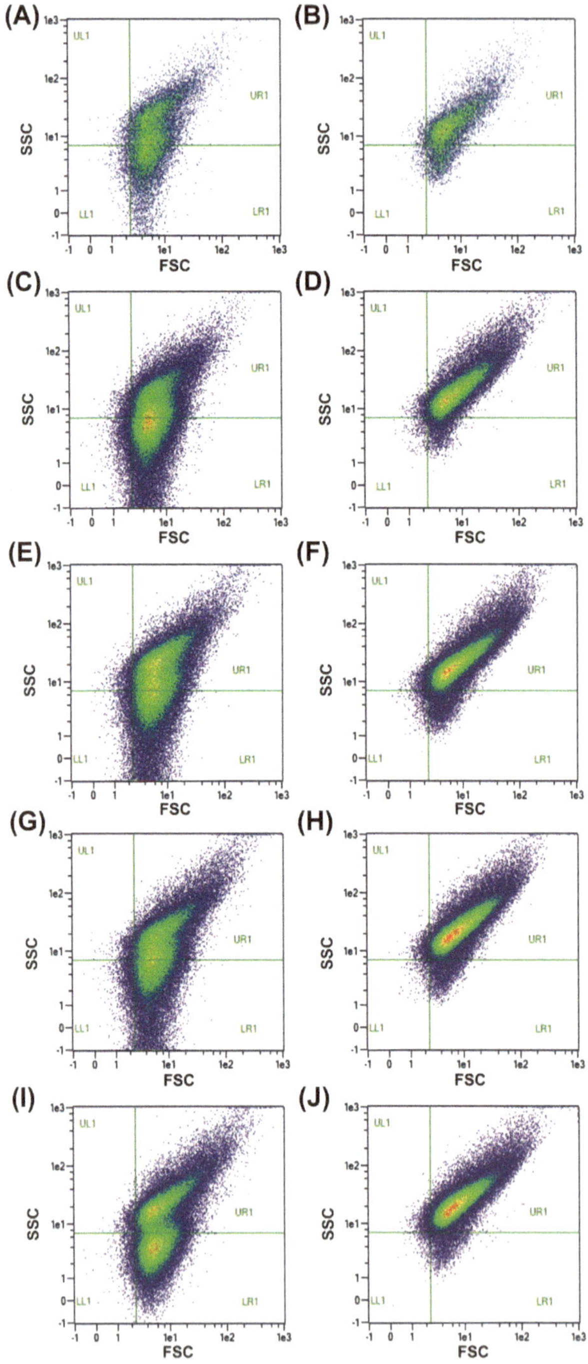

Fig. A5 Isolates from a frozen/thawed blood plasma measured on the day of preparation (left side panels, A,C,E,G, I) and the same samples that were frozen at -80 °C, thawed, and then stored for a week at 4 °C (right side panels; B, D, F, H, J). A,B, isolate #6; C, D, isolate #7; E, F, isolate #8; G, H, isolate #9; I, J, isolate #10.

Table A1 Absolute concentration of particles isolated from frozen/thawed plasma in the second experiment of centrifugation (at 8 °C).

Protocol	RCF	Duration	Temperature	Total number of isolated particles [a]
#6	10 000 g	60 min	8 °C	1.39E+05
#7	20 000 g	60 min	8 °C	2.68E+05
#8	50 000 g	60 min	8 °C	3.23E+05
#9	100 000 g	60 min	8 °C	2.21E+05
#10	100 000 g	120 min	8 °C	9.01E+04

RCF, relative centrifugal force.
[a]Calculated from concentration obtained by FCM and sample volumes.

Table A2 DLS results of isolates and their corresponding supernatants prepared from frozen/thawed plasma sample; obtained by protocols #6-#10.

isolate #6									
10 000g		dilution:	1:20						
Laser Intensity:		1.31E+01	mW						
Average Intensity I:		4.08E+01	kHz/mW						
Peaks		from t	to t	Mean t, s	Mean Rh, nm	mean G, s-1	mean D, cm2s-1	I, kHz/mW	I, % partcl <1000nm
	1	1.20E-05	1.20E-05	1.20E-05	1	8.33E+04	2.39E-06	0.05	1
	2	5.97E-05	5.97E-05	5.97E-05	5	1.68E+04	4.80E-07	0.03	2
	3	6.61E-04	8.64E-04	7.54E-04	64	1.33E+03	3.80E-08	5.93	3
	4	3.29E-03	5.61E-03	4.75E-03	407	2.10E+02	6.03E-09	23.55	4
	5	5.85E+00	5.85E+00	5.85E+00	>1000	1.71E-01	4.90E-12	11.27	5
						sum I. partcl Rh<1000nm		29.57	
						estimated total I, partcl Rh<1000nm		591.30	
Isolate #7									
20 000g		dilution:	1:20						
Laser Intensity:		4.60E+00	mW						
Average Intensity I:		6.85E+01	kHz/mW						
Peaks		from t	to t	Mean t, s	Mean Rh, nm	mean G, s-1	mean D, cm2s-1	I, kHz/mW	I, % partcl <1000nm
	1	1.20E-05	1.20E-05	1.20E-05	1	8.33E+04	2.39E-06	0.04	0%
	2	5.70E-04	7.15E-04	5.90E-04	51	1.70E+03	4.86E-08	7.42	23%
	3	2.79E-03	3.50E-03	3.07E-03	264	3.25E+02	9.33E-09	24.67	77%
	4	8.15E-01	8.15E-01	8.15E-01	>1000	1.23E+00	3.52E-11	36.36	
						sum I. partcl Rh<1000nm		32,13	
						estimated total I, partcl Rh<1000nm		642.57	
Isolate #8									
50 000g		dilution:	1:20						
Laser Intensity:		3.57E+00	mW						
Average Intensity I:		9.23E+01	kHz/mW						
Peaks		from t	to t	Mean t, s	Mean Rh, nm	mean G, s-1	mean D, cm2s-1	I, kHz/mW	I, % partcl <1000nm
	1	1.20E-05	1.20E-05	1.20E-05	1	8.33E+04	2.39E-06	0.05	0%
	2	4.65E-04	6.16E-04	5.60E-04	48	1.79E+03	5.12E-08	9.23	24%
	3	2.51E-03	3.33E-03	2.89E-03	249	3.46E+02	9.91E-09	29.55	76%
	4	3.16E-02	3.16E-02	3.16E-02	2717	3.16E+01	9.07E-10	8.53	
	5	1.16E+01	1.16E+01	1.16E+01	>1000	8.62E-02	2.47E-12	44.94	
						sum I. partcl Rh<1000nm		38.83	
						estimated total I, partcl Rh<1000nm		776.64	
Isolate #9									
100 000g		dilution:	1:10						
Laser Intensity:		2.05E+00	mW						
Average Intensity I:		1.64E+02	kHz/mW						
Peaks		from t	to t	Mean t, s	Mean Rh, nm	mean G, s-1	mean D, cm2s-1	I, kHz/mW	I, % partcl <1000nm
	1	1.40E-05	1.40E-05	1.40E-05	1	7.14E+04	2.05E-06	0.18	0%
	2	6.83E-05	6.83E-05	6.83E-05	6	1.46E+04	4.20E-07	0.25	0%
	3	4.33E-04	7.35E-04	5.90E-04	51	1.70E+03	4.86E-08	17.32	23%
	4	9.57E-04	3.58E-03	2.25E-03	193	4.44E+02	1.27E-08	56.92	76%
	5	6.07E-03	5.82E+00	3.58E+00	>1000	2.79E-01	8.00E-12	89.29	
						sum I. partcl Rh<1000nm		74.67	
						estimated total I, partcl Rh<1000nm		746.69	

(Continued)

Table A2 DLS results of isolates and their corresponding supernatants prepared from frozen/thawed plasma sample; obtained by protocols #6-#10.—cont'd

Isolate #10								
100 000g	dilution:	1:10						
Laser Intensity:	2.06E+00	mW						
Average Intensity I:	1.83E+02	kHz/mW						
Peaks	from t	to t	Mean t, s	Mean Rh, nm	mean G, s-1	mean D, cm2s-1	I, kHz/mW	I, % partcl <1000nm
1	1.20E-05	1.44E-05	1.27E-05	1	7.90E+04	2.27E-06	0.05	0%
2	2.08E-05	2.49E-05	2.32E-05	2	4.31E+04	1.23E-06	0.25	1%
3	1.08E-04	1.55E-04	1.42E-04	12	7.05E+03	2.02E-07	5.32	15%
4	6.70E-04	9.66E-04	8.29E-04	71	1.21E+03	3.46E-08	30.81	85%
5	3.47E-03	9.34E-02	4.15E-02	>1000	2.41E+01	6.91E-10	146.19	
						sum I. partcl Rh<1000nm	36.42	
						estimated total I, partcl Rh<1000nm	364.20	

supernatant #7								
20 000g	dilution:	1:20						
Laser Intensity:	4.60E+00	mW						
Average Intensity I:	6.85E+01	kHz/mW						
Peaks	from t	to t	Mean t, s	Mean Rh, nm	mean G, s-1	mean D, cm2s-1	I, kHz/mW	I, % partcl <1000nm
1	2.31E-05	3.21E-05	3.05E-05	3	3.28E+04	9.40E-07	2.19	1%
2	8.61E-05	1.66E-04	1.23E-04	11	8.10E+03	2.32E-07	16.83	7%
3	4.45E-04	3.19E-03	1.44E-03	123	6.95E+02	1.99E-08	219.74	92%
4	3.15E+01	3.15E+01	3.15E+01	>1000	3.17E-02	9.10E-13	15.61	
						sum I. partcl Rh<1000nm	238.76	
						estimated total I, partcl Rh<1000nm	4775.26	

supernatant #8								
50 000g	dilution:	1:20						
Laser Intensity:	1.32E+00	mW						
Average Intensity I:	2.53E+02	kHz/mW						
Peaks	from t	to t	Mean t, s	Mean Rh, nm	mean G, s-1	mean D, cm2s-1	I, kHz/mW	I, % partcl <1000nm
1	4.45E-05	6.47E-05	4.96E-05	4	2.02E+04	5.78E-07	7.78	3%
2	1.65E-04	3.49E-04	2.37E-04	20	4.21E+03	1.21E-07	36.67	15%
3	1.07E-03	2.27E-03	1.50E-03	129	6.65E+02	1.91E-08	198.59	83%
4	1.16E-01	1.16E-01	1.16E-01	>1000	8.62E+00	2.47E-10	10.39	
						sum I. partcl Rh<1000nm	243.03	
						estimated total I, partcl Rh<1000nm	4860.68	

supernatant #9								
100 000g	dilution:	1:20						
Laser Intensity:	1,29E+00	mW						
Average Intensity I:	2,55E+02	kHz/mW						
Peaks	from t	to t	Mean t, s	Mean Rh, nm	mean G, s-1	mean D, cm2s-1	I, kHz/mW	I, % partcl <1000nm
1	1.20E-05	1.20E-05	1.20E-05	1	8.33E+04	2.39E-06	0.12	0%
2	3.70E-05	4.90E-05	4.26E-05	4	2.35E+04	6.72E-07	3.58	1%
3	8.59E-05	2.00E-04	1.38E-04	12	7.27E+03	2.08E-07	15.21	6%
4	2.65E-04	4.41E-03	1.75E-03	151	5.70E+02	1.63E-08	233.61	93%
						sum I. partcl Rh<1000nm	252.51	
						estimated total I, partcl Rh<1000nm	5050.16	

supernatant #10								
100 000g	dilution:	1:20						
Laser Intensity:	1.33E+00	mW						
Average Intensity I:	2.61E+02	kHz/mW						
Peaks	from t	to t	Mean t, s	Mean Rh, nm	mean G, s-1	mean D, cm2s-1	I, kHz/mW	I, % partcl <1000nm
1	2.31E-05	3.21E-05	2.72E-05	2	3.68E+04	1.05E-06	1.34	1%
2	6.19E-05	1.66E-04	1.01E-04	9	9.95E+03	2.85E-07	15.85	6%
3	4.44E-04	6.13E-03	1.56E-03	135	6.39E+02	1.83E-08	227.42	93%
4	1.64E-02	3.17E-02	2.27E-02	1955	4.40E+01	1.26E-09	8.53	
5	8.48E-02	3.12E+01	1.96E+01	>1000	5.10E-02	1.46E-12	7.57	
						sum I. partcl Rh <1000nm	244.61	
						estimated total I. partcl Rh<1000nm	4892.29	

Table A3 Association of Alpha-2-macroglobulin (A2M), Apolipoprotein B-100 (APOB), C4b-binding protein alpha chain (C4BPA), Ceruloplasmin (CP), Complement C3 (C3), Complement C4-A (C4.A), Fibronectin (FN1), Immunoglobulin heavy constant gamma 2 (IGHG2), Immunoglobulin heavy constant mu (IGHM), Alpha-1-microglobulin/bikunin precursor (AMBP) and serum albumin (ALB) with membranes, EVs, and microparticles.

Annotation	Protein	Reference
blood microparticles	CP	PMID:22516433
	A2M	PMID:22516433
	C3	PMID:22516433
	IGHG2	PMID:22516433
	IGHM	PMID:22516433
	FN1	PMID:22516433
	AMBP	PMID:22516433
	ALB	PMID:22516433
	C4BPA	PMID:22516433
	C4A	PMID:22516433
	C4B	PMID:22516433
extracellular exosomes	CP	PMID:23533145. PMID:19056867
	A2M	PMID:23533145. PMID:20458337. PMID:21362503
	C3	PMID:23533145. PMID:21362503
	IGHG2	PMID:23533145
	IGHM	PMID:23533145. PMID:19199708
	FN1	PMID:23533145. PMID:21276792. PMID:19056867
	AMBP	PMID:23533145. PMID:19056867
	ALB	PMID:23533145. PMID:21276792. PMID:22516433. PMID:19056867. PMID:20458337. PMID:21362503
	APOB	PMID:23533145
	C4A	PMID:23533145
	C4B	PMID:23533145
citoplasmic membrane	CP	PMID:21873635
	C3	Reactome: R-HSA-977615. R-HSA-977371.R-HSA-977371. R-HSA-8853252. R-HSA-8852874. R-HSA-3266557
	IGHG2	GO_REF:0000037. GO_REF: 0000039
	IGHM	Reactome: R-HSA-983707. R-HSA-983703. R-HSA-983700. R-HSA-983696. R-HSA-9606151.

(*Continued*)

Table A3 Association of Alpha-2-macroglobulin (A2M), Apolipoprotein B-100 (APOB), C4b-binding protein alpha chain (C4BPA), Ceruloplasmin (CP), Complement C3 (C3), Complement C4-A (C4.A), Fibronectin (FN1), Immunoglobulin heavy constant gamma 2 (IGHG2), Immunoglobulin heavy constant mu (IGHM), Alpha-1-microglobulin/bikunin precursor (AMBP) and serum albumin (ALB) with membranes, EVs, and microparticles.—cont'd

Annotation	Protein	Reference
		R-HSA-8858498. R-HSA-5690740. R-HSA-5690702. R-HSA-5690701. R-HSA-1112666. GO_REF: 0000037. GO_REF:0000039
	AMBP	PMID:12817471. PMID:21873635
	C4BPA	Reactome: R-HSA-981680. R-HSA-981665. R-HSA-981658. R-HSA-981648. R-HSA-981637. R-HSA-977626
	APOB	Reactome: R-HSA-8871194. R-HSA-8871193. R-HSA-8869438. R-HSA-8868661. R-HSA-8868660. R-HSA-8868659. R-HSA-8868658. R-HSA-8868651. R-HSA-8868648. R-HSA-8868236. R-HSA-8868230. R-HSA-8868072. R-HSA-8868071. R-HSA-8867756. R-HSA-8867754. R-HSA-8863471. R-HSA-174706. R-HSA-174657. R-HSA-171141. R-HSA-171122
	C4A	Reactome:R-HSA-981637. R-HSA-977615
	C4B	Reactome:R-HSA-981637. R-HSA-977615
	FN1	GO_REF:0000107
	IGHG2	PMID:21873635
	IGHM	PMID:21873635
positive regulation of apoptotic cell clearance	C3	PMID:19302245
	C4A	PMID:19302245
	C4B	PMID:19302245

Acknowledgments

Authors acknowledge support from the European Union's Horizon 2020 research and innovation program under grant agreement No. 801338 (VES4US project) and Slovenian Research Agency (ARRS, grants J1-9162, J2-8166, J2-8169, J5-7098, P2-0132, P2-0232, P3-0388, Z2-9215).

References

[1] P. Wolf, The nature and significance of platelet products in human plasma, Br. J. Haematol. 13 (1967) 269–288.
[2] M. Yáñez-Mó, P.R.M. Siljander, Z. Andreu, et al., Biological properties of extracellular vesicles and their physiological functions, J. Extracell. Vesicles 4 (2015) 27066.
[3] D. Vozel, B. Uršič, J.L. Krek, R. Štukelj, V. Kralj-Iglič, Applicability of extracellular vesicles in clinical studies, Eur. J. Clin. Invest. 47 (2017) 305–313.
[4] R.E. Lane, D. Korbie, M.M. Hill, M. Trau, Extracellular vesicles as circulating cancer biomarkers: opportunities and challenges, Clin. Transl. Med. 7 (2018) 14.
[5] H. Shao, H. Im, X. Breakefield, R. Weissleder, H. Lee, New Technologies for analysis of extracellular vesicles, Chem. Rev. 118 (2018) 1917–1950.
[6] C. Campanella, C.C. Bavisotto, M. Logozzi, et al., On the choice of the extracellular vesicles for therapeutic purposes, Int. J. Mol. Sci. 20 (2019) E236.
[7] N. Kosaka, A. Kogure, T. Yamamoto, et al., Exploiting the message from cancer: the diagnostic value of extracellular vesicles for clinical applications, Exp. Mol. Med. 51 (2019) 31.
[8] O.P.B. Wiklander, M.Á. Brennan, J. Lötvall, X.O. Breakefield, S.E.l. Andaloussi, Advances in therapeutic applications of extracellular vesicles, Sci. Transl. Med. 11 (2019) eaav8521.
[9] E. Serrano-Pertierra, M. Oliveira-Rodriguez, M. Rivas, et al., Characterization of plasma-derived extracellular vesicles isolated by different methods: a comparison study, Bioengineering 6 (2019) E8.
[10] M.Y. Konoshenko, E.A. Lekchnov, A.V. Vlassov, P.P. Laktionov, Isolation of extracellular vesicles: general methodologies and latest trends, BioMed Res. Int. 2018 (2018) 8545347.
[11] T.A. Hartjes, S. Mytnyk, G.W. Jenster, V. van Steijn, M.E. van Royen, Extracellular vesicle quantification and characterization: common methods and emerging approaches, Bioengineering 6 (2019) E7.
[12] D.D. Taylor, S. Shah, Methods of isolating extracellular vesicles impact down-stream analyses of their cargoes, Methods 87 (2015) 3–10.
[13] M.I. Ramirez, M.G. Amorim, C. Gadelha, et al., Technical challenges of working with extracellular vesicles, Nanoscale 10 (2018) 881–906.
[14] K.W. Witwer, E.I. Buzas, L.T. Bemis, et al., Standardization of sample collection, isolation and analysis methods in extracellular vesicle research, J. Extracell. Vesicles 2 (2013).
[15] C. Théry, K.W. Witwer, E. Aikawa, et al., Minimal information for studies of extracellular vesicles 2018 (MISEV2018): a position statement of the International Society for Extracellular Vesicles and update of the MISEV2014 guidelines, J. Extracell. Vesicles 7 (2018) 1535750.
[16] M. Venturella, F.M. Carpi, D. Zocco, Standardization of blood collection and processing for the diagnostic use of extracellular vesicles, Curr. Pathobiol. Rep. 7 (2019) 1–8.
[17] V. Šuštar, A. Bedina-Zavec, R. Štukelj, et al., Nanoparticles isolated from blood: a reflection of vesiculability of blood cells during the isolation process, Int. J. Nanomed. 6 (2011) 2737–2748.

[18] M. Monguió-Tortajada, C. Gálvez-Montón, A. Bayes-Genis, S. Roura, F.E. Borràs, Extracellular vesicle isolation methods: rising impact of size-exclusion chromatography, Cell. Mol. Life Sci. 76 (2019) 2369–2382.
[19] T. Baranyai, K. Herczeg, Z. Onódi, et al., Isolation of exosomes from blood plasma: qualitative and quantitative comparison of ultracentrifugation and size exclusion chromatography methods, PLoS One 10 (2015) e0145686.
[20] L. Musante, M. Saraswat, A. Ravidà, B. Byrne, H. Holthofer, Recovery of urinary nanovesicles from ultracentrifugation supernatants, Nephrol. Dial. Transplant. 28 (2012) 1425–1433.
[21] Z. Onódi, et al., Isolation of high-purity extracellular vesicles by the combination of iodixanol density gradient ultracentrifugation and bind-elute chromatography from blood plasma, Front. Physiol. 9 (2018) 1479.
[22] N. Karimi, A. Cvjetkovic, S.C. Jang, R. Crescitelli, M.A. Hosseinpour Feizi, R. Nieuwland, J. Lötvall, C. Lässer, Detailed analysis of the plasma extracellular vesicle proteome after separation from lipoproteins, Cell. Mol. Life Sci. 75 (2018) 2873–2886.
[23] A. Abramowicz, P. Widlak, M. Pietrowska, Proteomic analysis of exosomal cargo: the challenge of high purity vesicle isolation, Mol. Biosyst. 12 (2016) 1407–1419.
[24] J. Van Deun, P. Mestdagh, R. Sormunen, et al., The impact of disparate isolation methods for extracellular vesicles on downstream RNA profiling, J. Extracell. Vesicles 3 (2014) 24858.
[25] R. Linares, S. Tan, C. Gounou, N. Arraud, A.R. Brisson, High-speed centrifugation induces aggregation of extracellular vesicles, J. Extracell. Vesicles 4 (2015) 29509.
[26] J.Z. Nordin, Y. Lee, P. Vader, et al., Ultrafiltration with size-exclusion liquid chromatography for high yield isolation of extracellular vesicles preserving intact biophysical and functional properties, Nanomedicine 11 (2015) 879–883.
[27] S.I. Brett, F. Lucien, C. Guo, et al., Immunoaffinity based methods are superior to kits for purification of prostate derived extracellular vesicles from plasma samples, Prostate 77 (2017) 1335–1343.
[28] P. Sharma, S. Ludwig, L. Muller, et al., Immunoaffinity-based isolation of melanoma cell-derived exosomes from plasma of patients with melanoma, J. Extracell. Vesicles 7 (2018) 1435138.
[29] A. Clayton, J. Court, H. Navabi, et al., Analysis of antigen presenting cell derived exosomes, based on immuno-magnetic isolation and flow cytometry, J. Immunol. Methods 247 (2001) 163–174.
[30] Y. Tian, M. Gong, Y. Hu, et al., Quality and efficiency assessment of six extracellular vesicle isolation methods by nano-flow cytometry, J. Extracell. Vesicles 9 (2020) 1697028.
[31] D. Freitas, M. Balmaña, J. Poças, et al., Different isolation approaches lead to diverse glycosylated extracellular vesicle populations, J. Extracell. Vesicles 8 (2019) 1621131.
[32] R.J. Berckmans, R. Neiuwland, A.N. Boeing, et al., Cell-derived microparticles circulate in healthy humans and support low grade thrombin generation, Thromb. Haemostasis 85 (2001) 639–646.
[33] M. Diamant, R. Nieuwland, R.F. Pablo, et al., Elevated numbers of tissue-factor exposing microparticles correlate with components of the metabolic syndrome in uncomplicated type 2 diabetes mellitus, Circulation 106 (2002) 2442–2447.
[34] C.A.R. Lok, A.N. Böing, P.H. Reitsma, et al., Expression of inflammation-related genes in endothelial cells is not directly affected by microparticles from preeclamptic patients, J. Lab. Clin. Med. 147 (2006) 310–320.
[35] B. Toth, K. Nikolajek, A. Rank, et al., Gender-specific and menstrual cycle dependent differences in circulating microparticles, Platelets 18 (2007) 515–521.

[36] R.J. Berckmans, R. Nieuwland, P.P. Tak, et al., Cell-derived microparticles in synovial fluid from inflamed arthritic joints support coagulation exclusively via a factor VII-dependent mechanism, Arthritis Rheum. 46 (2002) 2857–2866.
[37] R.J. Berckmans, R. Nieuwland, M.C. Kraan, et al., Synovial microparticles from arthritic patients modulate chemokine and cytokine release by synoviocytes, Arthritis Res. Ther. 7 (2005) R536–R544.
[38] R.J. Berckmans, A. Sturk, L.M. Van Tienen, M.C. Schaap, R. Nieuwland, Cell-derived vesicles exposing coagulant tissue factor in saliva, Blood 117 (2011) 3172–3180.
[39] I. Junkar, V. Utar, M. Frank-Bertoncelj, et al., Blood and synovial microparticles as revealed by atomic force and scanning electron microscope, Open Autoimmun. J. 1 (2009) 50–58.
[40] É. Biró, K.N. Sturk-Maquelin, G.M.T. Vogel, et al., Human cell-derived microparticles promote thrombus formation in vivo in a tissue factor-dependent manner, J. Thromb. Haemostasis 1 (2003) 2561–2568.
[41] E.A.J. Knijff-Dutmer, J. Koerts, R. Nieuwland, E.M. Kalsbeek-Batenburg, M.A.F.J. Van De Laar, Elevated levels of platelet microparticles are associated with disease activity in rheumatoid arthritis, Arthritis Rheum. 46 (2002) 1498–1503.
[42] K. Schara, R. Štukelj, J.L. Krek, et al., A study of extracellular vesicle concentration in active diabetic Charcot neuroarthropathy, Eur. J. Pharm. Sci. 98 (2017) 58–63.
[43] R. Štukelj, K. Schara, A. Bedina – Zavec, et al., Effect of shear stress in the flow through the sampling needle on concentration of nanovesicles isolated from blood, Eur. J. Pharm. Sci. 98 (2017) 17–29.
[44] A. Mrvar-Brečko, V. Šuštar, V. Janša, et al., Isolated microparticles from peripheral blood and body fluids as observed by scanning electron microscope, Blood Cells Mol. Dis. 44 (2010) 307–312.
[45] S. Liebhardt, N. Ditsch, R. Nieuwland, et al., CEA-, Her2/neu-, BCRP- and Hsp27-positive microparticles in breast cancer patients, Anticancer Res. 30 (2010) 1707–1712.
[46] B. Toth, R. Nieuwland, S. Liebhardt, et al., Circulating microparticles in breast cancer patients: a comparative analysis with established biomarkers, Anticancer Res. 28 (2008) 1107–1112.
[47] V. Lešer, D. Drobne, Ž. Pipan, M. Milani, F. Tatti, Comparison of different preparation methods of biological samples for FIB milling and SEM investigation, J. Microsc. 233 (2009) 309–319.
[48] S. Sitar, V. Aseyev, K. Kogej, Differences in association behavior of isotactic and atactic poly(methacrylic acid), Polymer 55 (2014) 848–854.
[49] S. Sitar, V. Aseyev, K. Kogej, Microgel-like aggregates of isotactic and atactic poly(methacrylic acid) chains in aqueous alkali chloride solutions as evidenced by light scattering, Soft Matter 10 (2014) 7712–7722.
[50] D. Božič, S. Sitar, I. Junkar, et al., Viscosity of plasma as a key factor in assessment of extracellular vesicles by light scattering, Cells 8 (2019) E1046.
[51] S.W. Provencher, CONTIN: a general purpose constrained regularization program for inverting noisy linear algebraic and integral equations, Comput. Phys. Commun. 27 (1982) 229–242.
[52] G. Pocsfalvi, L. Turiák, A. Ambrosone, et al., Protein biocargo of citrus fruit juice sac cells-derived vesicles reveals heterogeneous transport and extracellular vesicle subpopulations, J. Plant Physiol. 229 (2018) 111–121.
[53] C. Stanly, M. Moubarak, I. Fiume, L. Turiák, G. Pocsfalvi, Membrane transporters in citrus clementina fruit juice-derived nanovesicles, Int. J. Mol. Sci. 20 (2019) E6205.
[54] G. Pocsfalvi, C. Stanly, A. Vilasi, et al., Mass spectrometry of extracellular vesicles, Mass Spectrom. Rev. 35 (2016) 3–21.

[55] V. Kralj-Iglic, Stability of membranous nanostructures: a possible key mechanism in cancer progression, Int. J. Nanomed. 7 (2012) 3579–3596.
[56] S. Principe, E.E. Jones, Y. Kim, et al., In-depth proteomic analyses of exosomes isolated from expressed prostatic secretions in urine, Proteomics 13 (2013) 1667–1671.
[57] S.I. Buschow, B.W. van Balkom, M. Aalberts, et al., MHC class II-associated proteins in B-cell exosomes and potential functional implications for exosome biogenesis, Immunol. Cell Biol. 88 (2010) 851–856.
[58] W.D. Stamer, E.A. Hoffman, J.M. Luther, D.L. Hachey, K.L. Schey, Protein profile of exosomes from trabecular meshwork cells, J. Proteom. 74 (2011) 796–804.
[59] P. Bastos-Amador, F. Royo, E. Gonzalez, et al., Proteomic analysis of microvesicles from plasma of healthy donors reveals high individual variability, J. Proteomics 75 (2012) 3574–3584.
[60] P.A. Gonzales, T. Pisitkun, J.D. Hoffert, et al., Large-scale proteomics and phosphoproteomics of urinary exosomes, J. Am. Soc. Nephrol. 20 (2009) 363–379.
[61] S. Atay, C. Gercel-Taylor, M. Kesimer, D.D. Taylor, Morphologic and proteomic characterization of exosomes released by cultured extravillous trophoblast cells, Exp. Cell Res. 317 (2011) 1192–1202.
[62] M.P. Caby, D. Lankar, C. Vincendeau-Scherrer, G. Raposo, C. Bonnerot, Exosomal-like vesicles are present in human blood plasma, Int. Immunol. 17 (2005) 879–887.
[63] S. Strohkamp, T. Gemoll, J.K. Habermann, Possibilities and limitations of 2DE-based analyses for identifying low-abundant tumor markers in human serum and plasma, Proteomics 16 (2016) 2519–2532.
[64] D. Binns, E. Dimmer, R. Huntley, et al., QuickGO: a web-based tool for Gene Ontology searching, Bioinformatics 25 (2009) 3045–3046.
[65] E. Boilard, A.C. Duchez, A. Brisson, The diversity of platelet microparticles, Curr. Opin. Hematol. 22 (2015) 437–444.
[66] N. Arraud, R. Linares, S. Tan, et al., Extracellular vesicles from blood plasma: determination of their morphology, size, phenotype and concentration, J. Thromb. Haemostasis 12 (2014) 614–627.
[67] M.T. Aatonen, T. Öhman, T.A. Nyman, et al., Isolation and characterization of platelet-derived extracellular vesicles, J. Extracell. Vesicles 3 (2014) 24692.
[68] H.F. Heijnen, A.E. Schiel, R. Fijnheer, H.J. Geuze, J.J. Sixma, Activated platelets release two types of membrane vesicles: microvesicles by surface shedding and exosomes derived from exocytosis of multivesicular bodies and alpha-granules, Blood 94 (1999) 3791–3799.
[69] K. Joop, R.J. Berckmans, R. Nieuwland, et al., Microparticles from patients with multiple organ dysfunction syndrome and sepsis support coagulation through multiple mechanisms, Thromb. Haemostasis 85 (2001) 810–820.
[70] I. Melki, N. Tessandier, A. Zufferey, E. Boilard, Platelet microvesicles in health and disease, Platelets 28 (2017) 214–221.
[71] P.S. Sung, T.F. Huang, S.L. Hsieh, Extracellular vesicles from CLEC2-activated platelets enhance dengue virus-induced lethality via CLEC5A/TLR2, Nat. Commun. 10 (2019) 2402.
[72] B. Fendl, R. Weiss, M.B. Fischer, A. Spittler, V. Weber, Characterization of extracellular vesicles in whole blood: influence of pre-analytical parameters and visualization of vesicle-cell interactions using imaging flow cytometry, Biochem. Biophys. Res. Commun. 478 (2016) 168–173.
[73] A.P. Bode, C.L. Knupp, Effect of cold storage on platelet glycoprotein Ib and vesiculation, Transfusion 34 (1994) 690–696.
[74] J. Steinkühler, E. Sezgin, I. Urbančič, C. Eggeling, R. Dimova, Mechanical properties of plasma membrane vesicles correlate with lipid order, viscosity and cell density, Commun. Biol. 2 (2019) 337.

[75] S.L. Veatch, P. Cicuta, P. Sengupta, et al., Critical fluctuations in plasma membrane vesicles, ACS Chem. Biol. 3 (2008) 287–293.
[76] R. Zarka, M.B. Horev, T. Volberg, et al., Differential modulation of platelet adhesion and spreading by adhesive ligand density, Nano Lett. 19 (2019) 1418–1427.
[77] V. Leytin, Apoptosis in the anucleate platelet, Blood Rev. 26 (2012) 51–63.
[78] R. Nieuwland, R.J. Berckmans, R.C. Rotteveel-Eijkman, et al., Cell-derived microparticles generated in patients during cardiopulmonary bypass are highly procoagulant, Circulation 96 (1997) 3534–3541.
[79] A.J. Reininger, H.F. Heijnen, H. Schumann, et al., Mechanism of platelet adhesion to von Willebrand factor and microparticle formation under high shear stress, Blood 107 (2006) 3537.
[80] Y. Miyazaki, S. Nomura, T. Miyake, et al., High shear stress can initiate both platelet aggregation and shedding of procoagulant containing microparticles, Blood 88 (1996) 3456–3464.
[81] A. Pal, D.V. Khakhar, Breakage of vesicles in a simple shear flow, Soft Matter 15 (2019) 1979–1987.
[82] T. Nielsen, A.F. Kristensen, S. Pedersen, G. Christiansen, S.R. Kristensen, Investigation of procoagulant activity in extracellular vesicles isolated by differential ultracentrifugation, J. Extracell. Vesicles 7 (2018) 1454777.
[83] R.W. Stroetz, N.E. Vlahakis, B.J. Walters, M.A. Schroeder, R.D. Hubmayr, Validation of a new live cell strain system: characterization of plasma membrane stress failure, J. Appl. Physiol. 90 (2001) 2361–2370.
[84] O.E. Olorundare, O. Peyruchaud, R.M. Albrecht, D.F. Mosher, Assembly of a fibronectin matrix by adherent platelets stimulated by lysophosphatidic acid and other agonists, Blood 98 (2001) 117–124.
[85] W.S. To, K.S. Midwood, Plasma and cellular fibronectin: distinct and independent functions during tissue repair, Fibrogenesis Tissue Rep. 4 (2011) 21.
[86] T.T. Nguyen Huong, C. Huynh Khon, E. Scharf Rüdiger, R. Stoldt Volker, Biol. Chem. 394 (2013) 1495.
[87] H. Nguyen, K. Huynh, V.R. Stoldt, Shear-dependent fibrillogenesis of fibronectin: impact of platelet integrins and actin cytoskeleton, Biochem. Biophys. Res. Commun. 497 (2018) 797–803.
[88] E.I. Buzás, E.Á. Tóth, B.W. Sódar, K.É. Szabó-Taylor, Molecular interactions at the surface of extracellular vesicles, Semin. Immunopathol. 40 (2018) 453–464.
[89] G. Kushwaha, S.N. Chaurasia, A. Pandey, D. Dash, Characterization of fibrinogen binding on platelet-derived extracellular vesicles, Thromb. Res. 172 (2018) 135–138.
[90] S.P. Jackson, The growing complexity of platelet aggregation, Blood 109 (2007) 5087.
[91] I. Furi, F. Momen-Heravi, G. Szabo, Extracellular vesicle isolation: present and future, Ann. Transl. Med. 5 (2017), 263-263.
[92] E. Biró, R. Nieuwland, P.P. Tak, et al., Activated complement compounds and complement activator molecules on the surface of cell-derived microparticles in patients with rheumatoid arthritis and healthy individuals, Ann. Rheum. Dis. 66 (2007) 1085–1092.
[93] M.J. Justice, D.N. Petrusca, A.L. Rogozea, et al., Effects of lipid interactions on model vesicle engulfment by alveolar macrophages, Biophys. J. 106 (2014) 598–609.
[94] D. Božič, M. Hočevar, V. Kononenko, M. Jeran, U. Štibler, I. Fiume, M. Pajnič, L. Pađen, K. Kogej, D. Drobne, A. Iglič, G. Pocsfalvi, V. Kralj-Iglič, Effect of ultracentrifuge systemic force on extracellular vesicle isolates from blood plasma, (in preparation).

CHAPTER SIX

Multi-technique analysis of extracellular vesicles: not only size matters

Fabio Perissinotto[a], Beatrice Senigagliesi[a,b], Lisa Vaccari[a], Maria Pachetti[a,b], Francesco D'Amico[a], Heinz Amenitsch[c], Barbara Sartori[c], Karin Pachler[d,e], Magdalena Mayr[d], Mario Gimona[d,e], Eva Rohde[d,f], Federica Caponnetto[g], Daniela Cesselli[g], Loredana Casalis[a,*] and Pietro Parisse[a,*]

[a]Elettra Sincrotrone Trieste S.C.p.A., Trieste, Italy
[b]Università Degli Studi di Trieste, Trieste, Italy
[c]Graz University of Technology, Graz, Austria
[d]GMP Unit, Spinal Cord Injury and Tissue Regeneration Center Salzburg, Paracelsus Medical University, Salzburg, Austria
[e]Research Program "Nanovesicular Therapies", Paracelsus Medical University, Salzburg (Austria)
[f]Department of Transfusion Medicine, University Hospital, Salzburger Landeskliniken Ges.m.b.H (SALK), Salzburg, Austria
[g]Università Degli Studi di Udine, Udine, Italy
[*]Corresponding author: E-mails: pietro.parisse@elettra.eu, loredana.casalis@elettra.eu

Contents

Advances in Biomembranes and Lipid Self-Assembly, Volume 32
ISSN 2451-9634
https://doi.org/10.1016/bs.abl.2020.05.004

Abstract

Extracellular vesicles (EVs) are small vesicles ensuring transport of molecules between cells and throughout the body. Their small size and biological and physical functions make them optimal candidates as therapeutic agents in several medical frameworks. The elucidation of biophysical and biochemical characteristics of EVs is crucial for the understanding of their interaction with recipient cells and their functional activity. In particular, for therapeutic applications understanding how the process of manufacturing affects the biological function of EVs is mandatory before going to the clinical testing. The absence of standardized methodologies and technologies to establish reliable criteria has been the main hurdle for real therapeutic applications of EVs. Here we exploited a combination of different biophysical techniques (Fourier Transform Infrared Spectroscopy, Ultraviolet Resonant Raman Spectroscopy, Atomic Force Microscopy and Small Angle X-Ray Scattering) to screen EVs derived from umbilical cord-mesenchymal stem cells with different isolation procedures, evidencing subtle differences in the biophysical properties of EVs. The results underline the need of a multiparametric analysis to address size, stability and purity of different preparations and to correlate phenotypic parameters of EVs with their functional activity.

1. Introduction

Extracellular vesicles (EVs) are small vesicles released by almost all known cells ensuring the transport of molecules throughout the body [1]. According to the new definitions suggested by the International Society of Extracellular Vesicles, they can be divided in two main categories: small EVs, with sizes between 30 and 200 nm in diameter, mainly of endosomal origin, and large EVs, from sizes between 200 and 1000 nm in diameter, mainly originated from the outward budding of the cell membrane [2]. They are composed of a lipid bilayer enriched in adhesion molecules, ligands, receptors, membrane proteins and cell surface proteins responsible of their interaction with the recipient cells, that surrounds an internal cargo comprising nuclear proteins, ESCRT proteins, cytosolic proteins and nucleic acids [3]. Their cargo defines specific signatures and have been shown to strongly impact on the fate of recipient cells [4]. This effect, combined to the small size and natural biocompatibility, makes them perfect candidates as new generation cell-free therapeutic strategy [5,6], and in the latest years evidence of Mesenchymal Stem Cells (MSC) -derived EVs as therapeutic mediators in several frameworks (e.g. immune therapy, cell-free regenerative medicine, etc) has been demonstrated [7,8].

Although several observations of in vitro and in vivo potency have been reported, the therapeutic exploitation of MSC-EVs will require at least a basic understanding of the "Mode of Action" of EVs [9–11]. In fact the heterogeneity of the vesicles, due to their origin, makes it difficult to define standards for EV isolation and quantification, and a thorough analysis of the biophysical and biomolecular properties of EVs after isolation procedures is needed [12–14].

Here we will focus on the analysis of biophysical properties of standardized Umbilical Cord-MSC-EVs preparations, to help elucidating the role of phenotypic parameters (size, morphology, structure, protein/lipid ratio) in view of possible therapeutic applications. In particular we propose here a platform of unconventional techniques to capture morphological, structural and chemical information on EVs isolated with different protocols. We exploited the nanometer resolution of Atomic Force Microscopy to visualize the morphology of single EVs combined to structural information from Small Angle X-ray Scattering experiments and to the chemical information from vibrational spectroscopies, namely Fourier Transform Infrared and Ultraviolet Resonant Raman spectroscopies, to go beyond the mere size analysis and capture novel insights on the molecular contamination/stability of the different EVs preparations. This kind of approach should in our opinion be adopted any time an effect caused by EVs on distal cells in *vitro* or in vivo experiments needs to be ascertained.

2. Materials and methods

2.1 Isolation methods

We worked under Good Manufacturing Practice (GMP) conditions to expand multipotent MSCs derived from umbilical cord and then purified EVs by means of three different purification setups (Tangential Flow Filtration, Tangential Flow Filtration plus Ultracentrifugation, Tangential Flow Filtration plus Ultracentrifugation plus Size Exclusion Chromatography).

Tangential Flow Filtration (TFF) is a technique that allows the separation of particles within a fluid. It is based on the coupling of a permeable membrane filtration with a tangential flow and allows an efficient concentration of EVs from a colloidal matrix. At variance with conventional filtrations, due to the tangential flow with respect to the pores of the membrane, this technique is less sensitive to clogging effects due to large particles and therefore more suitable for large volume processing and in principle is scalable up to several liters [15]. Ultracentrifugation (UC) consists of a series of sequential steps of centrifugation at increasing speed to remove cell debris, aggregates of biopolymers, apoptotic bodies, and other particles with density higher than that of EVs, with a final step at >100,000 g [16]. The method needs optimization of procedures to avoid operator dependent yields, aggregation of vesicles, poor scalability and lack of purity, but it is still considered the gold standard technique in the field since the seminal work of Thery et al. [17]. Size Exclusion Chromatography (SEC) allows for the separation of particles of specific size through the elution in a column packed with a porous stationary phase that prevents the penetration of larger particles. The elution time of smaller particles is affected by the presence of the pores that increase their path causing them to elute after the large particles [16]. Even though the yield is usually lower with respect to other isolation procedures, EVs isolated through SEC have been shown to be highly pure and functional [18] and the combination of SEC with UC and/or TFF have been reported [19,20].

After isolation, the products have been processed, and stored at -80 °C under GMP compliant conditions before their successive biophysical, biochemical and functional characterization. It is worth noting that we maintained the same flow of production, reported in Fig. 1, for all the different set of purified EVs to limit the potential variability of results.

2.2 Manufacturing of human mesenchymal stromal cell–derived extracellular vesicles (EVs) and nanoparticle tracking analysis measurements

Human umbilical cord-derived mesenchymal stromal cells (UC-MSCs) were seeded (1560 cells/cm^2) into a CF4 cell factory system (Nunc) in fibrinogen-depleted culture medium composed of α-MEM (Sigma-Aldrich), 10% pHPL, and 5 mM (N2)-L-alanyl-L-glutamine (Fresenius Kabi). Cells were cultured at 5% CO_2 at 37 °C. Upon reaching a cell confluency of 60%–70%, cells were washed with PBS and the growth medium was exchanged to EV-harvest medium, consisting of α-MEM, 5% pHPL (which was additionally EV-depleted by tangential flow filtration (TFF) using a 750 kDa hollow fiber filter (Spectrum Labs)), and 5 mM (N2)-L-alanyl-L-glutamine. After 24 h, conditioned medium was harvested, centrifuged at 2.500 × g for 20 min at 18 °C, and filtered (0.22 μm). EVs were concentrated and buffer-exchanged to PBS by TFF using a 100 kDa hollow fiber filter (Spectrum Labs). Finally, EVs were isolated and concentrated by ultracentrifugation at 120.000 × g for 3 h at 18 °C. Resulting EV-pellets were washed with PBS and subsequently resuspended in Ringer's Lactate or in NaCl (for infrared spectroscopy experiments) in an appropriate volume to achieve a dose of 20–40 × 10^7/mL (cell equivalent). Resuspended EVs

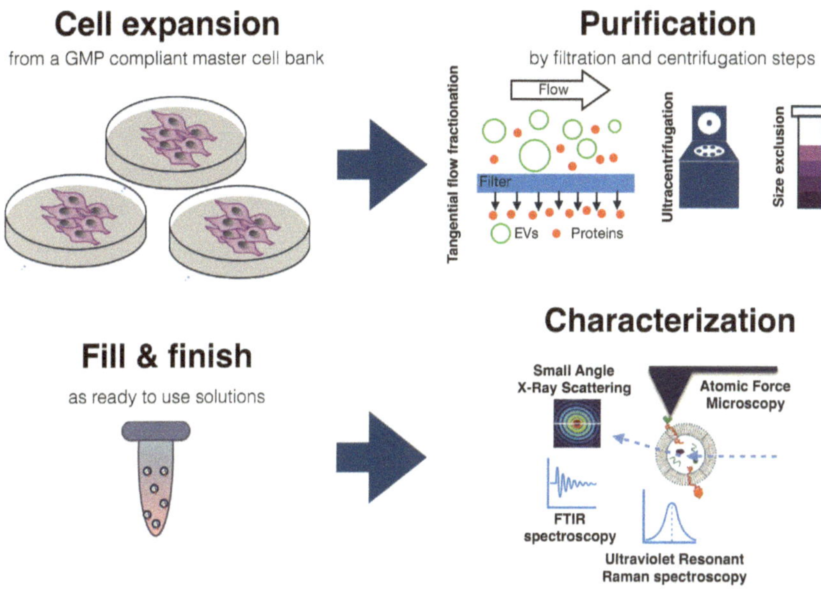

Fig. 1 Workflow for manufacturing EVs. After cell expansion from a GMP compliant master cell bank depleted of endogenous extracellular vesicles, EVs are isolated from conditioned medium. EV purification is performed using a combination of Tangential flow filtration, Ultracentrifugation and Size Exclusion Chromatography. Isolated and purified EVs are then stored in glass vials at −80 °C ready for quality control analyses. We kept as last step the characterization in order to avoid unwanted variables during the process.

were centrifuged at 3.000 × g for 10 min at 4 °C, sterile filtered (0.22 μm), and stored in glass vials at − 80 °C.

In certain cases, EVs were further purified by size exclusion chromatography (SEC) using qEVoriginal SEC columns (Izon Science) according to the manufacturer's instructions. All preparations were tested for the presence of typical EV/exosome markers, such as tetraspanins CD9, CD63 and CD81 in a Multiplex Bead- Based Flow cytometry assay (not reported here). EV concentrations were measured by means of Nanoparticle Tracking Analysis (NTA) using a ZetaView PMX 110 instrument (Particle Metrix) in light scatter mode. The minimum brightness was therefore set to 20 AU, temperature to 21.5 °C, shutter to 70 AU, and sensitivity to 85 AU. Subsequently, data for two exposures at 11 measurement positions were acquired per sample. Based on the Stokes-Einstein equation, particle sizes were determined using the ZetaView software (PMX 110: Version 8.4.2, PMX 120: Version 8.5.5).

2.3 Atomic force microscopy

Atomic Force Microscopy is a powerful scanning probe technique, capable of reconstructing the three-dimensional surface of samples with sub-nanometer resolution in x, y and z, exploiting the interaction force between a nanometer-sized tip and the sample itself. Since its invention in 1986 the technique has been applied to biological samples due to the possibility of measuring samples in liquid. Working in nearly physiological conditions at high spatial resolution makes the technique advantageous respect to electron microscopies, where special sample preparations are needed (staining, metallic plating, cryogenic procedures). The main drawback of AFM is the need of immobilization of the samples that can cause a partial loss of the free conformation of the biological molecules or objects in solution [21]. Nonetheless in the last 10 years the technique has been used for the characterization of Extracellular Vesicles providing useful insights on their shape and conformation [22–24] and where possible on their nano-mechanical properties [25–28]. The samples shown here were analyzed both in air and in liquid evidencing the information that can be retrieved in dry conditions and in solution.

Air measurements: A drop of 20 μL of solution was deposited on a freshly cleaved mica sheet for 15 min. Then we rinsed with milliQ water and dried with a gentle stream of nitrogen. The measurements have been performed in air with an NTMDT Solver Pro AFM in semi-contact mode with commercial silicon cantilevers (ETALON HA-NC, resonance frequency: 140–235 kHz, force constant: 3.5–12 N/m, Tip curvature radius < 10 nm). For a statistical analysis of the vesicle size we measured 5 regions of 10^2 μm^2 with a pixel size of 10 nm (comparable with tip radius) for all the EVs analyzed.

Liquid measurements: The measurements were performed in liquid with a Cypher AFM (Asylum Research – Oxford Instruments) in dynamic mode with commercial silicon nitride cantilevers (Olympus AC40TS, resonance frequency in liquid: 30 kHz, force constant: 0.1 N/m, Tip curvature radius < 10 nm). Also in this case a drop of 20 μL of solution was deposited on a freshly cleaved mica sheet for 15 min and then after removal of the initial solution we started imaging in phosphate buffered saline.

2.4 Small angle X-Ray scattering

Small Angle X-ray scattering is a powerful technique capable of providing structural information on objects in the sub-micrometer range. The technique is based on the elastic scattering of X-rays at the electrons of the investigated sample at small angles. The scattering intensity is measured as a function of the angle that is related to the momentum transfer q. The diffractogram of the scattering intensity versus the momentum transfer contains the reciprocal structural features of the sample of defined size d, as $q = 2\pi/d$. The structural information can be retrieved by fitting the measured data with theoretical models. SAXS is routinely applied for the study of nanoparticles and synthetic vesicular systems and has been recently used for the analysis of natural vesicles (synaptic vesicles [29] and EVs [30,31]).

The measurements were performed at the Austro-SAXS beamline at Elettra Sincrotrone Trieste [32], operated at a photon energy of 8 keV. The sample to detector distance was set to 1072 mm, resulting in a q-range within 0.1 and 5 nm^{-1}, as obtained from the calibration done with Silver behenate (d-spacing = 58.376 Å) [33]. The samples and relative buffers were dispensed by a custom made automatic sample changer [34] into an x-ray transparent, patented observation liquid cell [35]. The samples were thermalized to 25 °C and their temperature was controlled with a pt.100 probe positioned close to the observation cell. For each sample, 12 frames of 10 s were acquired with a Pilatus3 1 M detector, (Dectris, Baden, Switzerland), and the 2D images were radially averaged using the software Fit2D [36]. No radiation damage was detected. Correction for background, primary beam intensity and detector efficiency were included in the analysis of SAXS patterns.

2.5 Ultraviolet resonant raman spectroscopy

Raman spectroscopy is an inelastic light-scattering technique able to exploit the vibrational modes of molecules, which depend on their chemical conformation. Being the result of a second-order process, the diffused Raman radiation is characterized by a weak intensity. Thus, there is the need of enhancing the Raman signal. Here, we capitalize on the so-called Resonant Raman Spectroscopy: when the incident photon energy is close to an electronic transition of the material under examination, the Raman cross section is enhanced up to a factor 10^6–10^8.

In the particular case of biological systems, i.e. lipids, proteins and nucleic acids, it is possible to exploit the HOMO-LUMO transition in the UV and visible range to selectively enhance the specific Raman signals coming from specific chromophores and/or peculiar functional groups, even in complex systems.

Conventional Raman spectroscopy has been previously used for the analysis of EVs [37–39] and has been proven useful to detect differences in EVs preparations due both to isolation procedures and cell type [40]. By using UVRR spectroscopy instead we can selectively enhance the Raman signal coming from proteins and DNA with respect to the signal coming from the lipid and aliphatic components of the vesicles. The measurements taken on EVs were carried out at the UVRR setup installed at IUVS beamline (Elettra Sincrotrone Trieste) [41]. Spectra have been acquired at 226 nm wavelength for 15 min, on dried samples deposited on CaF_2 slides [40], with a continuous movement of the sample stage to reduce radiation damage effects.

2.6 Attenuated total reflectance-fourier transform infraRed measurements

Analogously and complementary to Raman spectroscopy, Infrared spectroscopy allows to investigate the covalent bonding pattern of samples by analyzing the first-order infrared (IR) absorptions due to the vibrational modes of constitutive molecules and molecular functional groups. Infrared measurements are routinely performed using the Fourier Transform InfraRed (FTIR) technique, that exploits an interferometric system to collect the desired spectrum. Intensity, position and bandwidth of the absorption bands allow to evaluate the chemical composition of a sample and the relative concentration of the different components of the sample itself. Being a label-free non-damaging technique, FTIR is widespread used for analysis of biological and pharmaceutical samples, to test, for example, their purity and reveal conformational changes of the biomolecules of interests. With respect to EV samples, the main drawback of FTIR spectroscopy is the need of working in dried conditions. The high absorption of water in the infrared spectral region would in fact hide the spectral features of interest. And even when working in dry conditions, FTIR measurements of thin samples as for EVs remain a challenge, making the measurements highly affected by the traces of the original buffer. To overcome the difficulties we chose to work in Attenuated Total Reflectance (ATR) geometry to enhance the surface contribution to the signal. Using multiple reflection crystals in fact, the signal-to noise ratio increases by a factor which is function of the number of reflections.

In the last years there have been few attempts of EVs FTIR measurements [42,43], able to distinguish characteristics spectral features and to evaluate the protein/lipid fraction in different EVs preparations. Starting from these works, we optimized the measurements by suspending our samples in sodium chloride buffer, which does not provide any vibrational feature overlapping with the spectra of interest, and acquiring spectra on thin dried EV films in ATR-FTIR geometry. The EV solution was let evaporate onto a diamond Internal Reflective Element (IRE) (9 reflections). Spectra were acquired at the Chemical and Life Sciences branch of SISSI (Synchrotron Infrared Source for Spectroscopy and Imaging, Elettra-Sincrotrone Trieste, Basovizza, Trieste, Italy), using the off-line Bruker Vertex 70 interferometer equipped with RT-DLaTGS detector. 30 consecutive spectra were acquired for monitoring the complete drying of the sample. Every spectrum was collected at 2 cm^{-1}, averaging 256 scans in the range 4500–800 cm^{-1}. The data-set was pre-processed with Bruker software OPUS 7.5. The spectra were first baseline corrected in the 4000–800 cm^{-1} spectra region (rubber band correction algorithm, 64 points) and then further baseline corrected, (rubber band method, 16 points) and normalized to the same height with a Min/Max function in order to better compare the band shapes among them and highlight the differences in their infrared features [44].

3. Results

3.1 Morphological analysis

We analyzed three sets of EVs derived from umbilical cord MSCs, prepared with different protocols of isolation (Tangential Flow Filtration, Tangential Flow Filtration plus Ultracentrifugation, Tangential Flow Filtration plus Ultracentrifugation plus Size Exclusion Chromatography) and analyzed by means of Nanoparticle Tracking Analysis (NTA), as described in Materials and Methods, immediately after isolation. In Table 1 we report the NTA results for the three preparations that evidence an increase of the average size measured and a reduction of the yield, increasing the purification steps from TFF to TFF + UC + SEC. The average size falls within the range of small EVs, therefore from hereafter we will refer to our preparations as sEVs.

In Fig. 2A–C, we report a typical AFM micrograph for the three sets of vesicles analyzed in air. The vesicles appear as roundish objects with, in some cases, a small indentation in the center due to the force exerted by the AFM tip [23]. The three samples observed present a different number of vesicles on the surface and also a different size distribution. In Fig. 2D and E we

Table 1 Samples analyzed with description of isolation method and concentration and average size measured by means of Nanoparticle Tracking Analysis.

Name	EV isolation method	NTA concentration (particles/mL)	NTA average size (nm)
sEV1	TFF 750 kD	$1–4 \times 10^{11}$	100.8
sEV2	TFF 750 kD + UC	$1–4 \times 10^{11}$	114
sEV3	TFF 750 kD + UC + SEC	$0.8–2 \times 10^{10}$	116

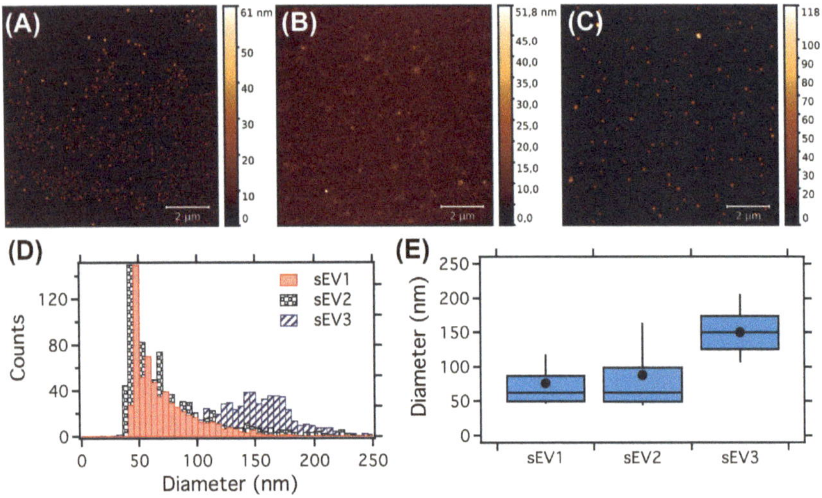

Fig. 2 (A–C) 10 µm × 10 µm Atomic Force Microscopy topographic maps acquired in air in semi-contact mode for three different sEV preparations (A) sEV1, TFF 750 kDa; (B) sEV2, TFF 750 kDa + UC; (C) sEV3, TFF 750 kDa + UC + SEC. (D–E) Histogram and box plot of diameter of sEVs extrapolated by the AFM images in (a, b, c). In the box plot (E) the *filled blue* area represents the population between 25th and 75th percentile. *The small bars* represents the 10th (lower) and the 90th percentile of the distribution. *Black filled circles* represent the average value.

report the histograms and the box plot analysis related to the diameter distribution of vesicles for sEV1, sEV2 and sEV3 samples. While sEV1 and sEV2 present a similar distribution, where most of the vesicles exhibit a diameter between 40 and 80 nm, sEV3 seems to be almost depleted in that region and have a quasi-normal distribution centered at around 150 nm. The average diameters of vesicles (reported as black dots in the box plot) extrapolated by AFM size distributions are 80 ± 30 nm for the sEV1, 90 ± 40 nm for sEV2, and 150 ± 40 nm for sEV3. If we compare the NTA and the AFM size analysis, we found that the most discordant result is for EV3. For EV1, EV2 results are similar, although NTA measures bigger average values. We need to point out here that the NTA technique is limited in reliably detecting particle' diameters below to 50 nm, due to the inherent limit in detection in the recording of scattered visible light: proteins and smaller aggregates are surely neglected by the NTA analysis, while are definitely measured by AFM. For the case of sEV3, the higher value measured with AFM might be due to the efficient removal of the protein content surrounding the EVs by the Size Exclusion Chromatography purification step [45], and to the selection of larger vesicles (the columns are optimized for 100 nm diameter particles). No relevant information can be retrieved on the height of the vesicles from the images acquired in air, since, as reported in Ref. [23] the drying process and the meniscus effect on the AFM tip can strongly perturb the measured height.

When moved to AFM measurements in physiological environment, we had a closer inspection on the shape of the particles obtained from different preparation conditions. sEV1 and sEV2 (Fig. 3 panels A and B) appear as roundish particles with dimensions comparable with the observations in air and show an average height ranging from 20 to 40 nm. We also observe an irregular background with a roughness of almost few nm, a sign for the presence of residual proteins on the mica surface. In the sEV3 sample (Fig. 3 panels C and D) instead the surface of mica seems to be cleaner (0.5 nm roughness), while the vesicles appear almost flat with a diameter of 80–100 nm and a height of 6–8 nm compatible with a lipid bilayer patch. This effect might be due to the opening and rupture of sEV3 on the surface, leaving a protein decorated lipid bilayer patch on the surface. The fact that this effect has been observed only with sEV3 could point to the higher instability of vesicles in the sEV3 preparations compared to sEV1 and sEV2. The presence of proteins in sEV1 and sEV2 preparations might act as a protecting agent that favors a soft absorption on the mica surface, that prevents the rupture of the vesicles.

3.2 Small angle X-ray scattering

In Fig. 4A we report the SAXS data acquired for the three different preparations. The q range used for the measurements allows to observe the presence of objects with a size ranging from 60 to few nanometers, that includes small protein aggregates, proteins and lipid bilayer thickness. We can observe that sEV1 and sEV2 samples present a similar scattering profile, except for the behavior at low q, while sEV3 sample is very different from the others.

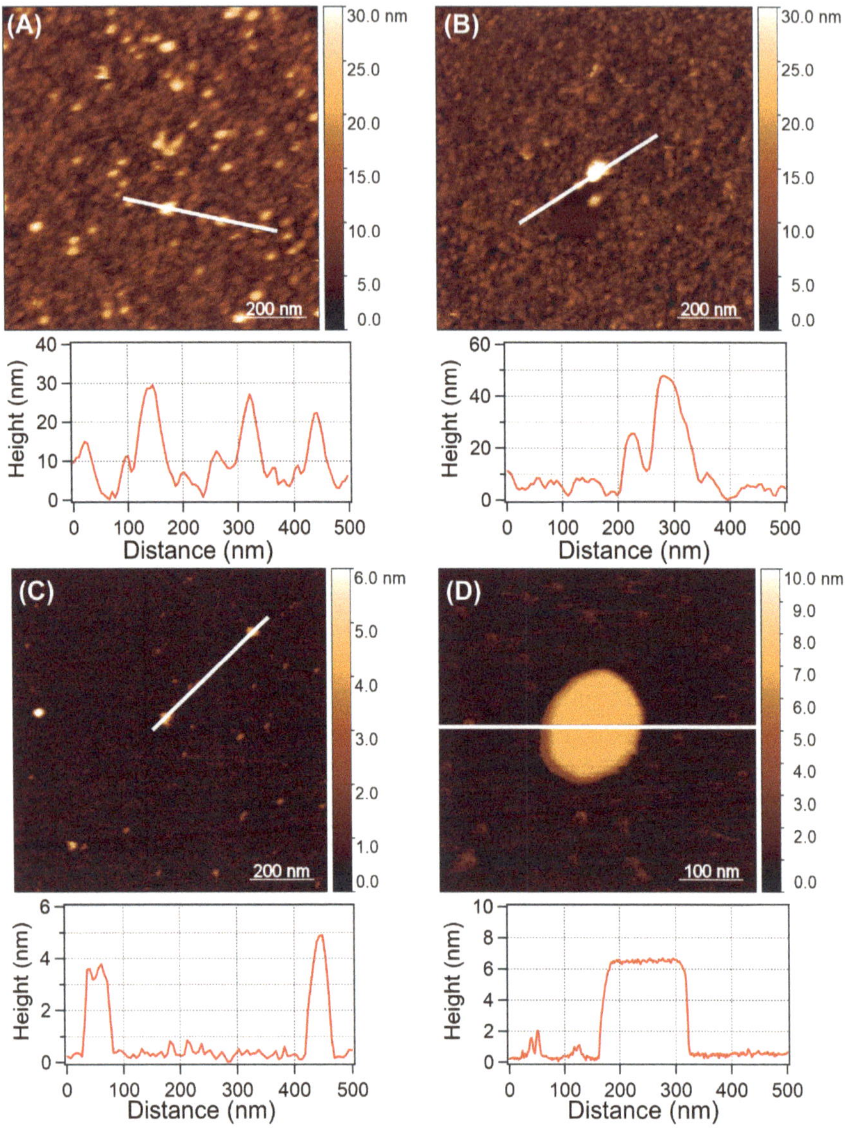

Fig. 3 (A–C) 1 μm × 1 μm Atomic Force Microscopy topographic maps acquired in liquid in semi-contact mode for three different sEV preparations and corresponding height profiles (A) sEV1, TFF 750 kDa; (B) sEV2, TFF 750 kDa + UC; (C) sEV3, TFF 750 kDa + UC + SEC. (D) 500 nm × 500 nm Atomic Force Microscopy topographic map of a single vesicle of sEV3 preparation and corresponding height profile.

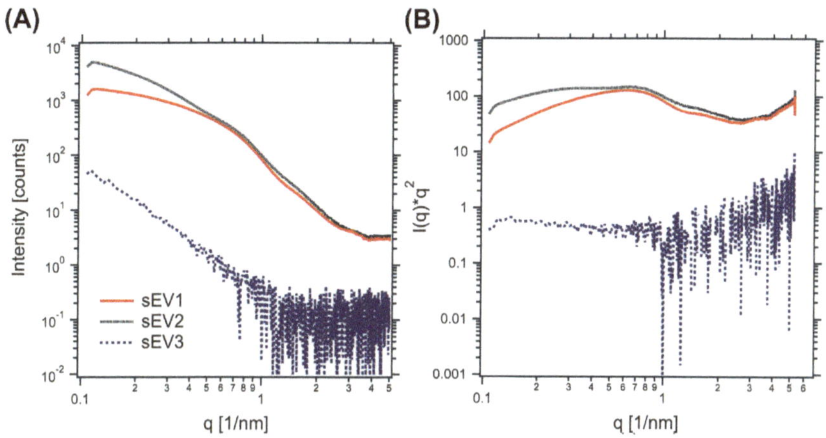

Fig. 4 Small Angle X-ray Scattering (panel A, Intensity and panel B, $I(q)*q^2$) measurements on the three different sEV preparations (*Continuous red line*: sEV1, TFF 750 kDa; *Dotted black line*: sEV2, TFF 750 kDa + UC; *Dashed blue line*: sEV3, TFF 750 kDa + UC + SEC).

Analyzing more in detail the shape of the scattering profile of sEV1 and sEV2 (enhanced in the $I(q)*q^2$ versus. q graph in Fig. 4B) we can clearly observe the presence of a bump around 0.6−1 nm^{-1} in both samples. The bump corresponds in real space to a size between 6 and 10 nm compatible with the dimensions of a protein-enriched bilayer. Therefore, it suggests the presence of the vesicles, as already reported by Romancino et al. [31]. The reduction of the scattering intensity of sEV1 at low q can be due to the presence of larger objects in the sEV2 preparation compared to the sEV1, as found by NTA analysis. Considering that with the ultracentrifuge we can partially remove proteins from the preparation, it is reasonable to observe the increase in the relative contribution of the vesicular content (30−150 nm) with respect to the protein content (5−20 nm in size) moving from sEV1 to sEV2. In the case of sEV3, despite the lower signal due to the 50−100 times lower yield of the sample prepared by size exclusion chromatography, we can observe a steep signal at low q indicating a distribution of large objects. The presence of the bump at 0.6−1 nm^{-1} is not visible in this case due to the low signal-to-noize ratio and could be influenced also by the reduced stability of sEV3 preparation, as observed by AFM analyses.

Qualitatively the SAXS data confirm the trend of an increased vesicular contribution at the expense of protein contribution, with the increase in the degree of purification. Moreover, the comparison of the scattering intensities of the different samples scales quite well with the concentrations measured with NTA, giving a relative concentration information, that could be useful for a fast screening of sample quality.

Unfortunately, the high level of inhomogeneity of our sEV samples makes it difficult to model the system (the previously developed method [29] relies on the strong approximation on the size distribution of vesicles and does not take into account the possibility of free proteins or small aggregates, present in our preparations) and to extract more quantitative information. Coupling SAXS with a size exclusion column directly on the beamline could facilitate the analysis of the different fractions and isolate the contributions of EVs of various sizes, proteins and protein aggregates, provided a sufficient concentration of the starting material.

3.3 Vibrational spectroscopies

In Figs. 5 and 6 we report the chemical characterization of the three preparations by ATR-FTIR and UV Resonant Raman spectroscopy, respectively. The analysis of the vibration spectra by FTIR and Raman reveals the presence of peaks attributed to lipids and proteins. In particular, the FTIR spectra are dominated by the presence of Amide I and II peaks at 1650 and 1540 cm^{-1} respectively. Differently, UVRR spectra are mainly characterized by the Amide I contribution, while Amide II is not clearly visible. By normalizing the spectra to the Amide I peak, we can clearly observe an evolution of the signatures of phospholipids (located at 2930–2850 cm^{-1}, 1739 cm^{-1} and 1235 cm^{-1} in FTIR spectra and around 2900 cm^{-1} in Raman ones) as a function of the different samples analyzed. In particular, the lipid band contribution increases from sEV1 to sEV2 with the maximum observed for sEV3. More in detail, in the IR spectra we can observe a decrease of the contribution of the band at 2873 cm^{-1} attributed to CH_3 symmetric stretching of proteins and an overall increase of the bands at 2960 cm^{-1} (CH_3 asymmetric stretching of lipids), 2933–2940 cm^{-1} (CH_2 asymmetric stretching of lipids), 2854 cm^{-1} (CH_2 symmetric stretching of lipids) and 1733–1739 cm^{-1} (C=O stretching attributed to ester functional groups in lipids) [46]. A similar trend is observed in the Raman spectra where we note an overall enhancement of the bands at 2933–2940 cm^{-1} (CH_3 asymmetric stretching of lipids) and 2892 cm^{-1} (CH_2 asymmetric stretching of lipids) [47]. In analogy to procedures reported in literature [42,43] for IR analysis of EVs' lipid/protein ratio, we extrapolated spectroscopic protein-to-lipid ratio from the FTIR and Raman spectra, analyzing the ratio between the protein peak areas and lipid peak areas. The values are reported in Table 2 and highlight the reduced protein

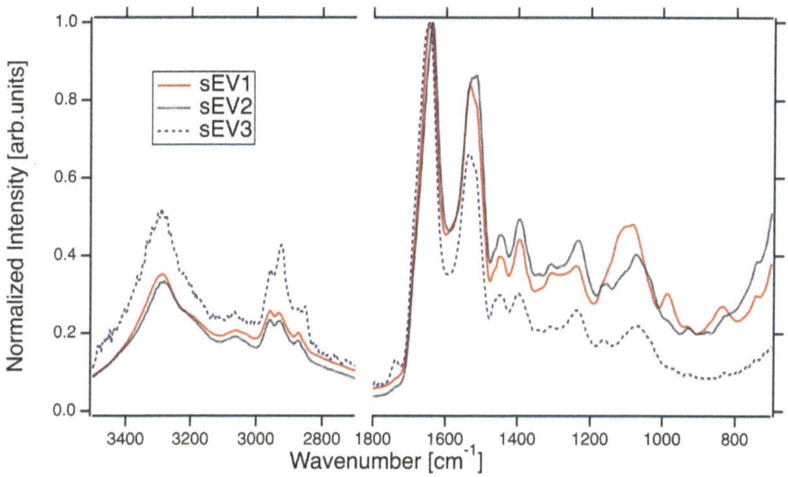

Fig. 5 Attenuated Total Reflectance -Infrared Spectroscopy, measurements on the three different EV preparations (*Continuous red line*: sEV1, TFF 750 kDa; *Dotted black line*: sEV2, TFF 750 kDa + UC; *Dashed blue line*: sEV3, TFF 750 kDa + UC + SEC). Spectra have been normalized the absorbance to the Amide I band to better compare the three different samples.

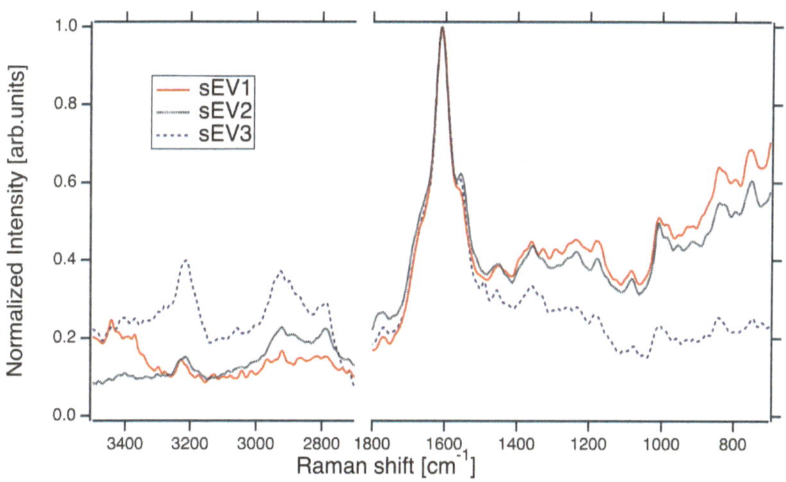

Fig. 6 UV Resonant Raman measurements on the three different EV preparations (*Continuous red line*: sEV1, TFF 750 kDa; *Dotted black line*: sEV2, TFF 750 kDa + UC; *Dashed blue line*: sEV3, TFF 750 kDa + UC + SEC). Spectra have been normalized to the absorbance to the Amide I band to better compare the three different samples.

Table 2 Protein/lipid ratio for the different EV preparations, as extracted by the analysis of Amide peaks and Phospholipids related peaks for ATR-FTIR and UVRR measurements.

Samples	FTIR protein/Lipid ratio	Raman protein/Lipid ratio
sEV1	14 ± 1	14 ± 1
sEV2	13 ± 1	4 ± 1
sEV3	6 ± 2	2.9 ± 0.7

content for sEV3 with respect to the other two preparations. The discrepancy between UVRR and FTIR extracted values in the case of sEV2 might be due to a difficulty in controlling the hydration state of the films during the drying process necessary for FTIR measurements, which in turn affects background removal. UVRR are instead more independent from water-related effects and, in this context, more trustable.

Apart from that, no other significant differences of peak positions and relative intensities have been observed in all EV samples, confirming an overall similar chemical composition of the three preparations.

4. Conclusions

In this chapter we reported the multiparametric analysis to evaluate the structural, morphological and chemical identity of EVs sorted in different ways. We tested highly reproducible vesicles produced in GMP conditions with three isolation protocols, with an increasing number of purification steps, namely Tangential Flow Filtration, Tangential Flow Filtration plus Ultracentrifuge and Tangential Flow Filtration plus Ultracentrifuge plus Size Exclusion Chromatography. Beside standard analytical tools, we used a platform of unconventional biophysical techniques as AFM, FTIR, UVRR and SAXS to get more molecular details on the different EVs preparations. Our analysis of EVs biophysical and biochemical properties confirms an increase of the ratio between vesicular content and total particle number going from sEV1 to sEV3, indicating that the addition of a step of size exclusion chromatography increases the efficiency in removing non-EV particles/protein content. The protein to lipid ratio trend revealed by both ATR-FTIR and UVRR further proves the effects of the purification protocols. Finally, the evolution of the vesicles size, as analyzed by AFM and NTA and suggested by the SAXS scattering patterns, foster these results, revealing the reduction of small aggregates when increasing the purification steps. Nonetheless, the highest purity of the vesicles after Size Exclusion

Chromatography has the downside of a lower yield that could hamper the use for therapeutic purposes, even though recent reports show the scalability to large volume samples [45,48,49]. Another issue of the size exclusion chromatography step is related to the vesicles' stability. As it appears from the AFM analysis in liquid, there is a higher extent of vesicles rupture in SEC treated samples, and this could be due to the absence of a stabilizing effect of the "free" proteins present in the other preparations. The hypothesis of a protein corona unspecifically surrounding the EVs in their native environment has been recently proposed and could also explain the observed instability [50]. To preserve EVs stability and functionality it might be possible to add stabilizers during storage, but this will require careful testing [51,52].

More in general, our results point toward the necessity of: (i) a multi parametric analysis, capable of giving a wide overview of the biophysical properties of EVs; (ii) monitoring the morphology of sEVs also in native environment, where AFM is almost unique among other microscopy technique; (iii) obtaining more information on the sample at the same time to evaluate not only size, but also purity and stability of the EV preparations.

While it could be difficult to access on standard basis to all the techniques proposed here, we highly recommend to combine different methodologies as in correlative microscopy. In particular coupling AFM high resolution morphology with nanomechanical measurements, or with functionalized tips a screening of mechanical and molecular fingerprints of subpopolations of EVs can be possible [27,28]. Also the coupling of label free techniques, such as interferometric imaging with AFM or with fluorescence microscopy can be envisaged [53,54]. From the biochemical information side, recently the coupling of infrared spectroscopy with scanning probe techniques enabled the realization of nano-Infrared and nano-Raman setups, capable of obtaining chemical maps with a few nanometer lateral resolution. The nanoIR approach has been recently applied for the study of Extracellular Vesicles and could open new routes in the understanding of EVs in terms of subpopulation identification and label free analysis [55,56].

Acknowledgments

The authors acknowledge funding from the European Regional Development Fund Interreg V-A Italia–Austria 2014–2020 (EXOTHERA ITAT1036). We acknowledge Elettra Sincrotrone Trieste for providing access to its synchrotron radiation facilities for the UVRR measurements at the IUVS beamline through proposal 20190210 and CERIC-ERIC Consortium for the access to experimental facilities at SAXS beamline and SISSI offline station through proposals 20187082 and 20172094. We thank Julien Lopez and Ted Limpoco from Asylum Research/Oxford Instruments for the experimental support in the Atomic Force Microscopy analyses in liquid.

References

[1] L. Margolis, Y. Sadovsky, The biology of extracellular vesicles: the known unknowns, PLoS Biol. 17 (2019) e3000363.

[2] C. Théry, K.W. Witwer, E. Aikawa, M.J. Alcaraz, J.D. Anderson, R. Andriantsitohaina, A. Antoniou, T. Arab, F. Archer, G.K. Atkin-Smith, D.C. Ayre, et al., Minimal information for studies of extracellular vesicles 2018 (MISEV2018): a position statement of the International Society for Extracellular Vesicles and update of the MISEV2014 guidelines, J. Extracell. Vesicles 7 (2018) 1.

[3] G. van Niel, G. D'Angelo, G. Raposo, Shedding light on the cell biology of extracellular vesicles, Nat. Rev. Mol. Cell Biol. 19 (2018) 213–228.

[4] M. Mathieu, L. Martin-Jaular, G. Lavieu, C. Thery, Specificities of secretion and uptake of exosomes and other extracellular vesicles for cell-to-cell communication, Nat. Cell Biol. 21 (2019) 9–17.

[5] S. Fais, L. O'Driscoll, F.E. Borras, E. Buzas, G. Camussi, F. Cappello, J. Carvalho, A. Cordeiro da Silva, H. Del Portillo, S. El Andaloussi, T. Ficko Trček, R. Furlan, A. Hendrix, I. Gursel, V. Kralj-Iglic, B. Kaeffer, M. Kosanovic, M.E. Lekka, G. Lipps, M. Logozzi, A. Marcilla, M. Sammar, A. Llorente, I. Nazarenko, C. Oliveira, G. Pocsfalvi, L. Rajendran, G. Raposo, E. Rohde, P. Siljander, G. van Niel, M.H. Vasconcelos, M. Yáñez-Mó, M.L. Yliperttula, N. Zarovni, A. Bedina Zavec, B. Giebel, Evidence-based clinical use of nanoscale extracellular vesicles in nanomedicine, ACS Nano. 10 (2016) 3886–3899.

[6] O.P.B. Wiklander, M.-Á. Brennan, J. Lötvall, X.O. Breakefield, El S. Andaloussi, Advances in therapeutic applications of extracellular vesicles, Sci. Transl. Med. 11 (2019) eaav8521.

[7] H. Valadi, K. Ekstrom, A. Bossios, M. Sjostrand, J.J. Lee, J.O. Lotvall, Exosome-mediated transfer of mRNAs and microRNAs is a novel mechanism of genetic exchange between cells, Nat. Cell Biol. 9 (2007) 654–659.

[8] T. Lener, M. Gimona, L. Aigner, V. Börger, E. Buzas, G. Camussi, N. Chaput, D. Chatterjee, F.A. Court, H.A. del Portillo, L. O'Driscoll, S. Fais, J.M. Falcon-Perez, U. Felderhoff-Mueser, L. Fraile, Y.S. Gho, A. Görgens, R.C. Gupta, A. Hendrix, D.M. Hermann, A.F. Hill, F. Hochberg, P.A. Horn, D. de Kleijn, L. Kordelas, B.W. Kramer, E.-M. Krämer-Albers, S. Laner-Plamberger, S. Laitinen, T. Leonardi, M.J. Lorenowicz, S.K. Lim, J. Lötvall, C.A. Maguire, A. Marcilla, I. Nazarenko, T. Ochiya, T. Patel, S. Pedersen, G. Pocsfalvi, S. Pluchino, P. Quesenberry, I.G. Reischl, F.J. Rivera, R. Sanzenbacher, K. Schallmoser, I. Slaper-Cortenbach, D. Strunk, T. Tonn, P. Vader, B.W.M. van Balkom, M. Wauben, S. El Andaloussi, C. Théry, E. Rohde, B. Giebel, Applying extracellular vesicles based therapeutics in clinical trials—an ISEV position paper, J. Extracell. Vesicles 4 (2015) 30087.

[9] K.W. Witwer, B.W.M. Van Balkom, S. Bruno, A. Choo, M. Dominici, M. Gimona, A.F. Hill, D. De Kleijn, M. Koh, R. Chai Lai, S.A. Mitsialis, L.A. Ortiz, E. Rohde, T. Asada, W. Seong Toh, D.J. Weiss, L. Zheng, B. Giebel, S.K. Lim, Defining mesenchymal stromal cell (MSC)-derived small extracellular vesicles for therapeutic applications, J. Extracell. Vesicles 8 (2019) 1609206.

[10] K.P. Robb, J.C. Fitzgerald, F. Barry, S. Viswanathan, Mesenchymal stromal cell therapy: progress in manufacturing and assessments of potency, Cytotherapy 21 (2019) 289–306.

[11] E. Rohde, K. Pachler, M. Gimona, Manufacturing and characterization of extracellular vesicles from umbilical cord–derived mesenchymal stromal cells for clinical testing, Cytotherapy 21 (2019) 581–592.

[12] L. Paolini, A. Zendrini, A. Radeghieri, Biophysical properties of extracellular vesicles in diagnostics, Biomarkers Med. 12 (2018) 383–391.
[13] Z. Khatun, A. Bhat, S. Sharma, A. Sharma, Elucidating diversity of exosomes: biophysical and molecular characterization methods, Nanomedicine 11 (2016) 2359–2377.
[14] N. Chopra, B. Dutt Arya, N. Jain, P. Yadav, S. Wajid, S.P. Singh, S. Choudhury, Biophysical characterization and drug delivery potential of exosomes from human wharton's jelly-derived mesenchymal stem cells, ACS Omega 4 (2019) 13143–13152.
[15] S. Busatto, G. Vilanilam, T. Ticer, W.-L. Lin, D. Dickson, S. Shapiro, P. Bergese, J. Wolfram, Tangential flow filtration for highly efficient concentration of extracellular vesicles from large volumes of fluid, Cells 7 (2018) 273.
[16] L.M. Doyle, M.Z. Wang, Overview of extracellular vesicles, their origin, composition, purpose, and methods for exosome isolation and analysis, Cells 8 (2019) 727.
[17] C. Théry, S. Amigorena, G. Raposo, A. Clayton, Isolation and characterization of exosomes from cell culture supernatants and biological fluids, Curr. Protocols Cell Biol. 30 (2006), 3.22.1-23.22.29.
[18] J.Z. Nordin, Y. Lee, P. Vader, I. Mäger, H.J. Johansson, W. Heusermann, O.P.B. Wiklander, M. Hällbrink, Y. Seow, J.J. Bultema, J. Gilthorpe, T. Davies, P.J. Fairchild, S. Gabrielsson, N.C. Meisner-Kober, J. Lehtiö, C.I.E. Smith, M.J.A. Wood, S.E.L. Andaloussi, Ultrafiltration with size-exclusion liquid chromatography for high yield isolation of extracellular vesicles preserving intact biophysical and functional properties, Nanomed. Nanotechnol. Biol. Med. 11 (2015) 879–883.
[19] A. Gámez-Valero, M. Monguió-Tortajada, L. Carreras-Planella, M. Franquesa, K. Beyer, F.E. Borràs, Size-Exclusion Chromatography-based isolation minimally alters Extracellular Vesicles' characteristics compared to precipitating agents, Sci. Rep. 6 (2016) 33641.
[20] J.Z. Nordin, R.B. Bostancioglu, G. Corso, S. EL Andaloussi, Tangential flow filtration with or without subsequent bind-elute size exclusion chromatography for purification of extracellular vesicles, in: J. Moll, S. Carotta (Eds.), Target Identification and Validation in Drug Discovery: Methods and Protocols, Methods in Molecular Biology, Humana Press, New York, NY, 1953.
[21] Y. Dufrêne, T. Ando, R. Garcia, D. Alsteens, D. Martinez-Martin, A. Engel, C. Gerber, D.J. Müller, Imaging modes of atomic force microscopy for application in molecular and cell biology, Nat. Nanotechnol. 12 (2017) 295–307.
[22] J. Hardij, F. Cecchet, A. Berquand, D. Gheldof, C. Chatelain, F. Mullier, B. Chatelain, J.-M. Dogné, Characterisation of tissue factor-bearing extracellular vesicles with AFM: comparison of air-tapping-mode AFM and liquid Peak Force AFM, J. Extracell. Vesicles 2 (2013) 1.
[23] P. Parisse, I. Rago, L. Ulloa Severino, F. Perissinotto, E. Ambrosetti, P. Paoletti, M. Ricci, A.P. Beltrami, D. Cesselli, L. Casalis, Atomic force microscopy analysis of extracellular vesicles, Eur. Biophys. J. 46 (2017) 813.
[24] N. Sebaihi, B. De Boeck, Y. Yuana, R. Nieuwland, J. Pétr, Dimensional characterization of extracellular vesicles using atomic force microscopy, Meas. Sci. Technol. 28 (2017) 034006.
[25] R. Sorkin, R. Huisjes, F. Bošković, D. Vorselen, S. Pignatelli, Y. Ofir-Birin, J.K. Freitas Leal, J. Schiller, D. Mullick, W.H. Roos, G. Bosman, N. Regev-Rudzki, R.M. Schiffelers, G.J.L. Wuite, Nanomechanics of extracellular vesicles reveals vesiculation pathways, Small 14 (2018) 1801650.
[26] D. Vorselen, S.M. van Dommelen, R. Sorkin, M.C. Piontek, J. Schiller, S.T. Döpp, S.A.A. Kooijmans, B.A. van Oirschot, B.A. Versluijs, M.B. Bierings, R. van Wijk, R.M. Schiffelers, G.J.L. Wuite, W.H. Roos, The fluid membrane determines mechanics of erythrocyte extracellular vesicles and is softened in hereditary spherocytosis, Nat. Commun. 9 (2018) 4960.

[27] S. Sharma, H.I. Rasool, V. Palanisamy, C. Mathisen, M. Schmidt, D.T. Wong, J.K. Gimzewski, Structural-mechanical characterization of nanoparticle exosomes in human saliva, using correlative AFM, FESEM, and force spectroscopy, ACS Nano 4 (2010) 1921–1926.
[28] B. Whitehead, L. Wu, M.L. Hvam, H. Aslan, M. Dong, L. Dyrskjøt, M.S. Ostenfeld, S.M. Moghimi, K.A. Howard, Tumour exosomes display differential mechanical and complement activation properties dependent on malignant state: implications in endothelial leakiness, J. Extracell. Vesicles 4 (2015) 29685.
[29] S. Castorph, D. Riedel, L. Arleth, M. Sztucki, R. Jahn, M. Holt, T. Salditt, Structure parameters of synaptic vesicles quantified by small-angle x-ray scattering, Biophys. J. 98 (2010) 1200–1208.
[30] Z. Varga, Y. Yuana, A.E. Grootemaat, E. van der Pol, C. Gollwitzer, M. Krumrey, R. Nieuwland, Towards traceable size determination of extracellular vesicles, J. Extracell. Vesicles 3 (2014) 1.
[31] D.P. Romancino, V. Buffa, S. Caruso, I. Ferrara, S. Raccosta, A. Notaro, Y. Campos, R. Noto, V. Martorana, A. Cupane, A. Giallongo, A. d'Azzo, M. Manno, A. Bongiovanni, Palmitoylation is a post-translational modification of Alix regulating the membrane organization of exosome-like small extracellular vesicles, Biochim. Biophys. Acta Gen. Subj. 1862 (2018) 2879–2887.
[32] H. Amenitsch, S. Bernstorff, P. Laggner, High-flux beamline for small-angle x-ray scattering at ELETTRA, Rev. Sci. Instrum. 66 (1995) 1624–1626.
[33] Y. Lu, Y. Yang, A. Sellinger, M. Lu, J. Huang, H. Fan, R. Haddad, G. Lopez, A.R. Burns, D.Y. Sasaki, J. Shelnutt, C.J. Brinker, Self-assembly of mesoscopically ordered chromatic polydiacetylene/silica nanocomposites, Nature 410 (2001) 913–917.
[34] NFFA deliverable Report D6.9: Nanodrop Cell for X-Ray and Nanoanalytic Applications.
[35] H. Amenitsch, B. Marmiroli, P. Laggner, US PATENT, 9164047, Bruker AXS GmbH, 2015.
[36] A.P. Hammersley, ESRF Internal Report, ESRF98HA01T, FIT2D V9.129 Reference Manual V3.1, 1998.
[37] A. Gualerzi, S. Niada, C. Giannasi, S. Picciolini, C. Morasso, R. Vanna, V. Rossella, M. Masserini, M. Bedoni, F. Ciceri, M.E. Bernardo, A.T. Brini, F. Gramatica, Raman spectroscopy uncovers biochemical tissue-related features of extracellular vesicles from mesenchymal stromal cells, Sci. Rep. 7 (2017) 9820.
[38] Z.J. Smith, C. Lee, T. Rojalin, R.P. Carney, S. Hazari, A. Knudson, K. Lam, H. Saari, E. Lazaro Ibañez, T. Viitala, T. Laaksonen, M. Yliperttula, S. Wachsmann-Hogiu, Single exosome study reveals subpopulations distributed among cell lines with variability related to membrane content, J. Extracell. Vesicles 4 (2015) 1.
[39] I. Tatischeff, E. Larquet, J.M. Falcón-Pérez, P.-Y. Turpin, S.G. Kruglik, Fast characterisation of cell-derived extracellular vesicles by nanoparticles tracking analysis, cryo-electron microscopy, and Raman tweezers microspectroscopy, J. Extracell. Vesicles 1 (2012) 1.
[40] A. Gualerzi, S.A.A. Kooijmans, S. Niada, S. Picciolini, A.T. Brini, G. Camussi, M. Bedoni, Raman spectroscopy as a quick tool to assess purity of extracellular vesicle preparations and predict their functionality, J. Extracell. Vesicles 8 (2019) 1.
[41] F. D'Amico, M. Saito, F. Bencivenga, M. Marsi, A. Gessini, G. Camisasca, E. Principi, R. Cucini, S. Di Fonzo, A. Battistoni, E. Giangrisostomi, C. Masciovecchio, UV resonant Raman scattering facility at Elettra, Nucl. Instrum. Methods Phys. Res. Sect. A Accel. Spectrom. Detect. Assoc. Equip. 703 (2013) 33–37.
[42] J. Mihály, R. Deák, I.C. Szigyártó, A. Bóta, T. Beke-Somfai, Z. Varga, Characterization of extracellular vesicles by IR spectroscopy: fast and simple classification based on amide and CH stretching vibrations, Biochim. Biophys. Acta 1859 (2017) 459–466.

[43] T. Visnovitz, X. Osteikoetxea, B.W. Sódar, J. Mihály, P. Lőrincz, K.V. Vukman, E.Á. Tóth, A. Koncz, I. Székács, R. Horváth, Z. Varga, E.I. Buzás, An improved 96 well plate format lipid quantification assay for standardisation of experiments with extracellular vesicles, J. Extracell. Vesicles 8 (2019) 1.
[44] F. Perissinotto, C. Stani, E. De Cecco, L. Vaccari, V. Rondelli, P. Posocco, D. Scaini, G. Legname, P. Parisse, L. Casalis, Iron-Mediated interaction of alpha synuclein with raft-like model membranes, Nanoscale 12 (2020) 7631.
[45] G. Corso, I. Mäger, Y. Lee, A. Görgens, J. Bultema, B. Giebel, M.J.A. Wood, J.Z. Nordin, S. El Andaloussi, Reproducible and scalable purification of extracellular vesicles using combined bind-elute and size exclusion chromatography, Sci. Rep. 7 (2017) 11561.
[46] N. Toyran, B. Turan, F. Severcan, Selenium alters the lipid content and protein profile of rat heart: an FTIR microspectroscopic study, Arch. Biochem. Biophys. 458 (2007) 184–193.
[47] K. Czamara, K. Majzner, M. Pilarczyk, K. Kochan, A. Kaczor, M. Baranska, Raman spectroscopy of lipids: a review, J. Raman Spectrosc. 46 (2014) 4–20.
[48] K. Takov, D.M. Yellon, S.M. Davidson, Comparison of small extracellular vesicles isolated from plasma by ultracentrifugation or size exclusion chromatography: yield, purity and functional potential, J. Extracell. Vesicles 8 (2019) 1560809.
[49] R.E. Lane, D. Korbie, M. Trau, M.M. Hill, Optimizing size exclusion chromatography for extracellular vesicle enrichment and proteomic analysis from clinically relevant samples, Proteomics 19 (2019) e1800156.
[50] M. Skliar, V.S. Chernyshev, D.M. Belnap, G.V. Sergey, S.M. Al-Hakami, P.S. Bernard, I.J. Stijleman, R. Rachamadugu, Membrane proteins significantly restrict exosome mobility, Biochem. Biophys. Res. Commun. 501 (2018) 1055–1059.
[51] A. Jeyaram, S.M. Jay, Preservation and storage stability of extracellular vesicles for therapeutic applications, AAPS J. 20 (2018) 1.
[52] M. Richter, K. Fuhrmann, G. Fuhrmann, Evaluation of the storage stability of extracellular vesicles, JoVE 147 (2019) e59584.
[53] G. Daaboul, P. Gagni, L. Benussi, P. Bettotti, M. Ciani, M. Cretich, D.S. Freedman, R. Ghidoni, A.Y. Ozkumur, C. Piotto, D. Prosperi, B. Santini, M.S. Ünlü, M. Chiari, Digital detection of exosomes by interferometric imaging, Sci. Rep. 6 (2016) 37246.
[54] A. Gori, A. Romanato, G. Bergamaschi, A. Strada, P. Gagni, R. Frigerio, D. Brambilla, R. Vago, S. Galbiati, S. Picciolini, M. Bedoni, G.G. Daaboul, M. Chiari, M. Cretich, Membrane-binding peptides for extracellular vesicles on-chip analysis, J. Extracell. Vesicles 9 (2020) 1.
[55] S.Y. Kim, D. Khanal, P. Tharkar, B. Kalionis, W. Chrzanowski, None of us is the same as all of us: resolving the heterogeneity of extracellular vesicles using single-vesicle, nanoscale characterization with resonance enhanced atomic force microscope infrared spectroscopy (AFM-IR), Nanoscale Horiz. 3 (2018) 430–438.
[56] S.Y. Kim, D. Khanal, B. Kalionis, W. Chrzanowski, High-fidelity probing of the structure and heterogeneity of extracellular vesicles by resonance-enhanced atomic force microscopy infrared spectroscopy, Nat. Protoc. 14 (2019) 576–593.

Kolibri 360 GmbH
Ferdinand-Jühlke-Straße 7
99095 Erfurt, Deutschland
produktsicherheit@kolibri360.de